Lecture Notes in Computer Science 829
Edited by G. Goos and J. Hartmanis

Andrew Chmora Stephen B. Wicker (Eds.)

Error Control, Cryptology, and Speech Compression

Workshop on Information Protection
Moscow, Russia, December 6-9, 1993
Selected Papers

Springer-Verlag
Berlin Heidelberg New York
London Paris Tokyo
Hong Kong Barcelona
Budapest

Series Editors

Gerhard Goos
Universität Karlsruhe
Postfach 69 80
Vincenz-Priessnitz-Straße 1
D-76131 Karlsruhe, Germany

Juris Hartmanis
Cornell University
Department of Computer Science
4130 Upson Hall
Ithaca, NY 14853, USA

Volume Editors

Andrew Chmora
Institute for Information Transmission Problems
19 Yermolovoy St., Moscow 101447 GSP-4 Russia

Stephen B. Wicker
Georgia Institute of Technology, School of Electrical and Computer Engineering
Atlanta, Georgia 30332, USA

CR Subject Classification (1991): E.3-4

ISBN 3-540-58265-7 Springer-Verlag Berlin Heidelberg New York
ISBN 0-387-58265-7 Springer-Verlag New York Berlin Heidelberg

CIP data applied for

Printed in Germany

Typesetting: Camera-ready by author
SPIN: 10472623 45/3140-543210 - Printed on acid-free paper

Preface

This book contains selected papers from the Workshop on Information Protection held in Moscow, Russia, on December 6–9, 1993. The Workshop gathered nearly 50 participants. This was possible due to the generous support and enthusiastic encouragement of the Russian Chapter of the IEEE Information Theory Society and the Institute for Information Transmission Problems of the Russian Academy of Sciences. This support is gratefully acknowledged by the Workshop Organizing Committee and the Editors of this volume.

This Workshop was the first open conference in Russia devoted mostly to cryptography and authentication. However, we note that the users of communication systems need various forms of protection for their transmitted data. Therefore, another important direction covered by the Workshop was the protection of information from errors. Another reason for this choice of topics is that both cryptology and error-control coding often employ common theoretical tools. The Organizing Committee hopes that workshops devoted to this joint research area will become traditional.

The papers have been revised and updated by the authors and editors. They are organized in three sections: Cryptology, Error-Control Coding, and Speech Compression.

The papers in the first section are concerned with the following problems of cryptology: algebraic aspects of key generation systems, the susceptibility of digital signature schemes based on error-correcting codes to universal forgery, message protection in cryptosystems modelled as the generalized wire-tap channel II, the avoidance of the Sidel'nikov-Shestakov attack, and a linear algebraic approach to secret sharing schemes.

The second section is devoted to error-control codes. Among the problems covered in this volume are generalizations of the Griesmer bound, coverings for decoding by S-sets, the periodicity of one-dimensional tilings, codes that can correct two-dimensional bursts, self-checking decoding algorithms for Reed-Solomon codes, an interesting construction of unit memory and partial unit memory codes including optimal nonlinear codes, a construction of concatenated codes based on convolutional codes as outer and inner codes, and the investigation of trellis coded modulation for nonlinear channels with intersymbol interference.

The final section is concerned with digital transmission of speech. The first paper considers the problem of reducing the number of multiplications in infinite response filtering. The second paper demonstrates a method for using trellis codes in linear predictive speech compression.

The editors are most grateful to all the authors for their hard work in preparing, presenting, and revising their papers. Without the dedication and enthusiasm of the authors the Workshop itself would not have been possible, let alone this book. Special thanks should be expressed to Professor Stephen B. Wicker, whose intensive support made the appearance of this volume possible.

Victor Zyablov, Chairman of Organizing Committee.

Table of Contents

Cryptology

Error Control Coding

Speech Compression

Algebraic Aspects of Key Generation Systems

V.A.Artamonov, A.A.Klyachko, V.M.Sidelnikov, V.V.Yashchenko

Moscow State University

Abstract. The principle of organization of private communication using public channels was suggested in 1976 by W. Diffie and M. E. Hellman [1]. After their publication, variants of the realization of such private communication were invited. Comparing known systems, we identify three general mathematical objectives among these variants:

1. the collection of sets and maps which are used by each subscriber for various computations;
2. a protocol describing information exchange;
3. a mathematical problem whose complexity provides for the security of the key generation system.

In this paper we consider the first objective in its simplest form: a protocol that includes only one round of information exchange.

1 Key Generation Using Universal Algebras of a Special Sort

The algebraic model for a one-round key generation protocol is the following collection of 5 sets and 4 maps with an identity relation:
K_1, K_2 - 2 sets of secret keys for 2 subscribers;
U_1, U_2 - 2 sets of public keys for 2 subscribers;
S - the set of common keys
$\phi_1 : K_1 \to U_1, \phi_2 : K_2 \to U_2, \psi_1 : K_1 \times U_2 \to S, \psi_2 : K_1 \times U_2 \to S$ with the identity

$$\psi_1(x, \phi_2(y)) = \psi_2(x, \phi_1(y))$$

We call the following collection a *pk*-algebra:

$$A = A(K_1, K_2, U_1, U_2, S, \phi_1, \phi_2, \psi_1, \psi_2)$$

The key-generation protocol for two subscribers A_1, A_2 is the following:

1. each subscriber A_i chooses a secret key $k_i \in K_i$;
2. each subscriber A_i calculates a public key $u_i = \phi_i(k_i) \in U_i$;
3. subscribers exchange their u_i;
4. subscriber A_i calculates $s_i = \psi_i(k_i, u_{3-i}) \in S$, which in fact does not depend on i.

The security of the system is based on the complexity of the following problem. **Problem 1.** Given u_1 and u_2, find s $(= s_i)$.
It is clear that s depends only on u_i i.e., there exists a function $f : U_1 \times U_2 \to S$,

such that $f(\phi_1(k_1), \phi_2(k_2))$ is equal to $s = s(k_1, k_2)$, calculated by the above protocol . Thus to solve problem 1 it is sufficient to solve the following problem.
Problem 2. Given u_i find a solution to the equation

$$\phi_i(x) = u_i$$

2 Elementary Properties of *pk*-Algebras

It is evident that the class of all *pk*-algebras is a variety of 5-base universal algebras. Thus we can speak about homomorphisms, subalgebras, congruences, factor algebras and direct products.
Consider two examples of *pk*-algebras.

Diffie-Hellman system [1]. Let $U_1 = U_2 = S = F_q^*$ be the multiplicative group of the finite field of q elements, ξ - a fixed element of $F_q^*; K_1 = K_2 = \{1, 2, \ldots, q-1\}, \phi_1(k) = \phi_2(k) = \xi^k, \psi_1(k, f) = \psi_2(k, f) = f^k$.

Noncommutative-semigroup-system [2]. Let H and R be some abelian subsemigroups of a noncommutative semigroup G.

$$K_1 = K_2 = H \times R$$
$$U_1 = U_2 = S = G$$
$$\phi_1(h, r) = hr = \phi_2(h, r)$$
$$\psi_1(h, r, g) = \psi_2(h, r, g) = hrg$$

The main identity holds due to the commutativity of the subsemigroups.

Consider the following simple class of *pk*-algebras which includes both examples. Let

$$K_1 = K_2 = K, U_1 = U_2 = S,$$
$$\phi_1(k) = \phi_2(k) = \rho(k, a)$$
$$\psi_1 = \psi_2 = \rho,$$

where $\rho : K \times S \to S$ is a fixed map, and $a \in S$ is a fixed element. The main identity is equivalent to the following:

$$\rho(k_1, \rho(k_2, a)) = \rho(k_2(\rho(k_1, a)) \tag{1}$$

Note that if K is a subset of a commutative semigroup L whose action on the set S is defined by the map ρ, then the main identity holds.

The following theorem shows that all the solutions of this functional equation can be obtained in this way.

Theorem 1. *Let $\rho_a : K \to Sk \mapsto \rho(k, a)$ be a bijective map. Then the map ρ is a solution of (1) iff K is a commutative groupoid with a neutral element e and the multiplication is related with ρ by the identity*

$$\rho(k_1, \rho(k_2, a)) = \rho(k_1 k_2, a)$$

and the neutral element is $\rho_a^{-1}(a)$.

Proof. Sufficiency is evident. Let $\rho : K \to S$ be a solution of (1). The following statement holds:

$$\forall k_1 \forall k_2 \exists! k_3 \rho(k_1, \rho(k_2, a)) = \rho(k_3, a)$$

Thus we define $k_1 k_2 = k_3$. Obviously this operation satisfies all the conditions. The theorem is proved.

Such algebras are convenient for realization. The map ρ can be implemented by means of some finite automaton. Each subscriber possessing this automaton can execute all the operations for key generation. Problem 2 in this case takes the following revised form.

Problem 3. Given an element $s \in S$, find an unknown element x such that

$$\rho(x, a) = s$$

This problem is a generalization of the well-known discrete logarithm problem.

3 A Functional Description of *pk*-Algebras

Consider a triple of sets X, Y, Z and a map $f : X \to Y$. Let $K_1' = X$, $U_1' = Y$, $K_2' = Z^Y$ (Z^Y is the set of all functions $Y \to Z$), $U_2' = Z^X$, $S' = Z$, $\phi_1' = f$, $\phi_2' : g \mapsto g \circ f$, $\psi_1'(x, h) = h(x)$, $\psi_2'(h, y) = h(y)$. It is evident that $(K_1', K_2', U_1', U_2', S')$ is a *pk*-algebra, say $F(X, Y, Z, f)$.

Theorem 2. *The class* **F** *of all pk-algebras defined above is universal in the following sense: for any pk-algebra B there exists an algebra A in* **F** *and a homomorphism $B \to A(B)$ identical on the sets S, U_1 and K_1.*

Proof. Consider an arbitrary *pk*-algebra $B = (K_1, K_2, U_1, U_2, S)$, algebra $A(B) = F(K_1, U_1, S, \phi_1) \in \mathbf{F}$. Define a morphism $\Phi : B \to A$ as identical on the sets K_1, U_1, S and state $\Phi(k)(y) = \psi_2(k, y), k \in K_2, y \in U_1; \Phi(u)(x) = \psi_1(x, u), u \in U_2, x \in K_1$. It is evident that Φ is a homomorphism of *pk*-algebras.

This description is in some sense sufficiently explicit; however, one may ask the following question. Is it possible to construct such a natural class *pk*-algebra **G** so that any *pk*-algebra is embedded by a natural way in some algebra from the class **G** ?

One can make this question more explicit in the following way. Does there exist for a given category **C** (in this paper $\mathbf{C} = PK$ - the variety of all *pk*-algebras) a more simple category **K** and correspondences $F : \mathbf{K} \to \mathbf{C}$, $G : \mathbf{C} \to \mathbf{K}$ and $H : \mathbf{C} \to IMor\mathbf{C}$ ($IMor\mathbf{C}$ is the category of injective morphisms of category **C**), such that the following diagram in the category of categories is commutative. Here α and ω are the standard functors "beginning" and "end", and Id is the identical functor.

In the case when such a triple of natural maps exists we will say that the category **C** is injectively interpreted in the category **K**.

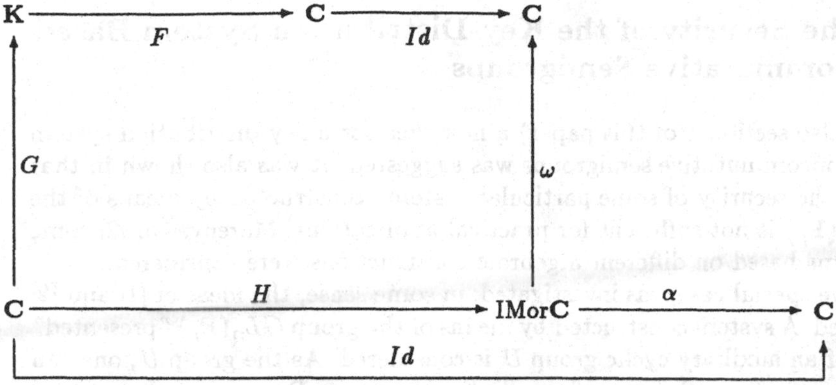

An example of such an interpretation is given by the Calley theorem about the embedding of any group in a permutation group. In this case **C** is the category of groups, **K** - the category of sets, G the forgetting functor (i.e. the functor mapping group G to the set G), F establishes the correspondence of the set M to the permutation group of M, and H establishes the correspondence of the group G to the natural injections to the permutation group of the set G. Generally we can say that if some variety of universal algebras is interpreted in the category of all universal algebras of some signature, then identities of the variety are not artificial in some sense.

Let category **K** be the category of all 5-base algebras with 3 operations $(X, Y, Z, T, P, f : X \to Y, \psi : T \times Y \to Z, \phi : T \to P)$.

Define $F : \mathbf{K} \to PK, G : PK \to \mathbf{K}, H : PK \to$IMor(PK) in the following way:

$F(X, Y, Z, T, P, f, \psi, \phi) = (K_1 = X, K_2 = T, U_1 = Y, U_2 = Z^X \times P, S = Z, \psi_1, \psi_2 = \psi, \phi_1 = f, \phi_2 = f \times \phi)$, where $\psi_1(x, g, p) = g(x)$.

$G(K_1, K_2, U_1, U_2, S, \psi_1, \psi_2, \phi_1, \phi_2) = (X = K_1, Y = U_1, Z = S, T = K_2, P = U_2, f = \phi_1, \psi = \psi_2, \phi = \phi_2)$

$H(K_1, K_2, U_1, U_2, S, \psi_1, \psi_2, \phi_1, \phi_2) = \gamma : (K_1, K_2, U_1, U_2, S, \psi_1, \psi_2, \phi_1, \phi_2) \to (K_1, K_2, U_1, (S^K 1) \times U_2, S, \tau, \psi_2, \phi_1, \phi_2)$, where $\tau(k, g, u) = g(k)$, $\gamma(id, id, id, \rho, id)$, $\rho = (\delta \times id) : U_2 \to (S^K 1) \times U_2$, and $\delta : U_2 \to S^{K1}$ is the map obtained in the natural way from ψ_1.

This construction satisfies all the conditions, therefore the following theorem is proved.

Theorem 3. *The category of pk-algebras is injectively interpreted in the category of all 5-base algebras with 3 operations:* $(X, Y, Z, T, P, f : X \to Y, \psi : T \times Y \to Z, \phi : T \to P)$

4 On the Security of the Key-Distribution System Based on Noncommutative Semigroups

In [4] (see also section 1 of this paper) a new idea for a key-distribution system based on noncommutative semigroups was suggested. It was also shown in that paper that the security of some particular systems constructed by means of the group $GL_n(\mathbf{F}_p)$ is not sufficient for practical applications. Moreover in [2] some other systems based on different algebraic constructions were considered.

In [3] one special case was investigated; in some sense, the ideas of [1] and [2] are combined. A system constructed by means of the group $GL_n(\mathbf{F}_p)$ "presented" by means of an auxiliary cyclic group U is considered. As the group U, one can take as an example a subgroup of an elliptic curve over $\mathbf{F}_q$.

In [3] the case $U =< \eta >$, a subgroup of order p of the multiplicative group of an auxiliary field $\mathbf{F}_q, p \mid q-1$, and G, the group of affine transformations of the field $\mathbf{F}_p, G \subset GL_2(\mathbf{F}_p)$, is considered. In this case the problem of obtaining the common key s by an adversary under some additional conditions is equivalent to the following having known elements $\eta, \eta^x, \eta^y, \eta^z$ to find $\eta^{xy/z}$. This problem is not equivalent to the Diffie-Hellman problem (that of obtaining η^{xy} having known η^x and η^y).

If one takes $G = GL_2(\mathbf{F}_p)$ some new parameters appear which apparently increase security.

References

1. W.Diffie and M. Hellman, New Directions in Cryptography. *IEEE Trans. Inform. Theory* **22** (1976), 472-492.
2. V.M.Sidelnikov, M.A.Cherepniov, W.W.Yaschenko. The key distribution system based on the use of noncommutative groups. (In Russian) Doklady RAN, v. 332, *N*.5, 1993.
3. V.M.Sidelnikov. A key generation system based on the exponentiation presentation of the linear group $GL_n(\mathbf{F}_p)$. Problemy Peredachi Informatcii (to appear)
4. V.M.Sidelnikov. A key generation system based exponentiation and noncommutative group. *IEEE Intern. Symp. on Inform. Theory.* 27 June-1 July 1994, Trondheim Norway.

Susceptibility of Digital Signature Schemes Based on Error-Correcting Codes to Universal Forgery

Mohssen Alabbadi and Stephen B. Wicker

Coding and Information Theory Laboratory
School of Electrical and Computer Engineering
Georgia Institute of Technology
Atlanta, Georgia 30332 USA
+1 (404) 894-3129 (Voice)
+1 (404) 853-9959 (FAX)
wicker@ee.gatech.edu

Abstract. Xinmei's digital signature scheme and the scheme's modified version as proposed by Harn and Wang have been shown by the authors and others to be susceptible to several different attacks. The authors have since devised and presented a scheme that is impervious to the attacks that were successfully applied to the earlier schemes. It is shown in this paper that this new scheme and Xinmei's scheme are vulnerable to universal forgeries. Equipped with this attack and the earlier ones, general remarks about digital signature schemes based on linear error-correcting block codes are presented.

1 Introduction

In 1990, Xinmei presented a true trapdoor digital signature scheme based on linear error-correcting block codes. The scheme was later modified by Harn and Wang to reduce the threat of selective forgery. Both schemes were subsequently shown by the authors and others to be susceptible to a variety of attacks. The authors then devised a scheme that is impervious to the attacks that were successful on the previous schemes. It is shown in this paper that this new scheme as well as Xinmei's scheme are vulnerable to universal forgeries. Equipped with this attack and the previous ones, general remarks about digital signature schemes based on linear error-correcting block codes are concluded. These remarks may be used as guidelines to construct a secure scheme. The next two sections contain brief reviews of Xinmei's and the authors' digital signature schemes. This is followed by a discussion of two efficient attacks that result in universal forgery for both schemes. The final section sets out several requirements for a truly secure digital signature scheme based on linear block error correcting codes.

2 Xinmei's Digital Signature Scheme

In Xinmei's digital signature scheme [13], each user, say user A, chooses an (n, k) binary Goppa code C_A that has the ability to correct t_A errors. A $k \times n$ binary generator matrix G_A and an $(n-k) \times n$ binary parity check matrix H_A are then selected for C_A. User A finds an $n \times k$ binary matrix G^* such that $G_A G_A^* = I_k$, where I_k is the $k \times k$ identity matrix. User A selects two nonsingular binary matrices: an $n \times n$ matrix P_A and a $k \times k$ matrix S_A, then he/she computes the following matrices:

$$J_A = P_A^{-1} G_A^* S_A^{-1} \tag{1}$$
$$W_A = G_A^* S_A^{-1} \tag{2}$$
$$T_A = P_A^{-1} H_A^T. \tag{3}$$

User A publishes J_A, W_A, T_A, H_A, t_A, and t'_A where $t'_A < t_A$, but $S_A G_A$ and P_A constitute the private key.

User A obtains the n-bit signature $\underline{c}_j$ of the k-bit message $\underline{m}_j$ by computing

$$\underline{c}_j = (\underline{e}_j \oplus \underline{m}_j S_A G_A) P_A, \tag{4}$$

where $\underline{e}_j$ is a random n-bit error vector of Hamming weight $w_H(\underline{e}_j) = t'_A < t_A$. The receiver validates the signature $\underline{c}_j$ by applying the Berlekamp-Massey algorithm on the syndrome $\underline{c}_j T_A = \underline{e}_j H_A^T$ to obtain $\underline{e}_j$, which must have weight of t'_A. Then J_A, W_A, and $\underline{e}_j$ are used to recover $\underline{m}_j$ by computing the expression

$$\underline{m}_j = \underline{c}_j J_A \oplus \underline{e}_j W_A. \tag{5}$$

In [1] the linearity of the code and knowledge of the error vectors are exploited in a chosen-message attack that results in a total break of Xinmei's scheme. The attack transforms the cryptanalytic problem into a pair of systems of linear equations: one containing n equations in n variables, and the other containing k equations in k variables. The complexity of this attack is thus $O(n^3)$.

It was observed by Harn and Wang in [5] that the combination of valid signatures of some messages yields a valid signature for another message; Xinmei's scheme is thus vulnerable to selective forgeries. Harn and Wang proposed a modification of Xinmei's scheme that appears to secure it against selective forgery. Their scheme has been shown to be totally breakable under known-message attack [2]. In [12] van Tilburg devised a direct attack that totally breaks both the Xinmei scheme and the Harn-Wang modified version of Xinmei's scheme. Under such attack the private key is directly obtained from the public key.

3 The Authors' Scheme

The authors have presented a scheme [3] that overcomes the weaknesses of Xinmei's scheme and the Harn-Wang scheme. In the authors' scheme, each user, say user A, selects an (n, k) binary irreducible Goppa code C_A that has the ability to correct t_A errors. User A then selects a $k \times n$ binary generator matrix G_A and an $(n-k) \times n$ binary parity check matrix H_A for the code C_A. The user then finds G_A^* such that $G_A G_A^* = I_k$, where I_k is the $k \times k$ identity matrix. A nonsingular binary $n \times n$ matrix P_A is then generated, and the matrices $G'_A = P_A^{-1} G^*_A$ and $H'_A = P_A^{-1} H_A^T$ are computed. Finally, user A selects an $n \times l$ binary matrix R_A of rank n, where $n < l$, and determines R^*_A such that $W_A W^*_A = I_n$. The public key consists of G'_A, H'_A, H_A, R^*_A, t_A, and t'_A, where t'_A is an integer such that $t'_A < t_A$. The private key consists of the matrices G_A, P_A, G^*_A, and R_A. Furthermore, a nonlinear noninvertible function $f(\underline{x}, \underline{y})$ is made available to all users, where $\underline{x}$ is a binary k-tuple, $\underline{y}$ is a binary n-tuple, and the output value is a binary k-tuple. The function f can be implemented in a similar fashion to the DES [10].

A k-bit message $\underline{m}_j$ is signed in the following manner. A random binary error vector $\underline{z}_j$ of length n and weight t'_A is selected. A random l-bit vector $\underline{e}_j$ of arbitrary non-zero weight is also selected. The l-bit signature $\underline{s}_j$ is then computed using the expression

$$\underline{s}_j = [(\underline{z}_j \oplus [f(\underline{m}_j, \underline{z}_j) \oplus \underline{z}_j G^*_A] G_A) P_A \oplus \underline{e}_j R^*_A] R_A \oplus \underline{e}_j. \tag{6}$$

$\underline{s}_j$ and $\underline{m}_j$ are transmitted. The signature is validated by first computing

$$\underline{v}_j = \underline{s}_j R^*_A = (\underline{z}_j \oplus [f(\underline{m}_j, \underline{z}_j) \oplus \underline{z}_j G^*_A] G_A) P_A. \tag{7}$$

The Berlekamp-Massey algorithm is then applied to the syndrome $\underline{v}_j H'_A = \underline{z}_j H_A^T$ to obtain $\underline{z}_j$, which must have weight of t'_A. Then G'_A is used to recover $f(\underline{m}_j, \underline{z}_j)$ as $\underline{v}_j G'_A$. Finally $\underline{v}_j G'_A$ is compared with the hashing function value $f(\underline{m}_j, \underline{z}_j)$ obtained using the received $\underline{m}_j$ and the computed $\underline{z}_j$. The signature is accepted if the two are identical.

The scheme is impervious to the attacks that are successful on Xinmei's scheme and the Harn-Wang scheme. However, this scheme as well as the previous ones are vulnerable to universal forgery as will be shown in the next section. We will first show that the matrix R_A has no cryptographic significance and the problem is reduced to generating an n-bit vector $\underline{v}_j$ that is accepted by the validation process, for there exits an $n \times l$ binary matrix R' such that $R' R^*_A = I_n$ and $\underline{s}_j$ is then obtained as $\underline{s}_j = \underline{v}_j R'$. The matrix R' can be found in polynomial time as follows. n linearly independent rows of W^*_A are selected (this can be done by row reduction of the matrix W^*_A, requiring $O(ln^2)$ bit operations). Let the n linearly independent rows be numbered as $l_1, l_2, \cdots, l_n$. The n linearly independent rows are then inverted in $O(n^3)$ bit operations. The columns of the inverted matrix correspond to columns $l_1, l_2, \cdots, l_n$ of R' and the other $l - n$ columns of R' are filled with zeros.

4 Universal Forgery

4.1 Attack I

Knowledge of the error vectors alone could jeopardize the security of the schemes; a known-message attack is devised that allows the cryptanalyst to universally forge signatures. We begin by noting that, in Xinmei's scheme,

$$\underline{c}_j = (\underline{e}_j \oplus \underline{m}_j S_A G_A) P_A = \underline{e}'_j \oplus \underline{m}_j E_A, \tag{8}$$

where $\underline{e}'_j = \underline{e}_j P_A$ and $E_A = S_A G_A P_A$. Similarly, in the other scheme,

$$\underline{v}_j = (\underline{z}_j \oplus [f(\underline{m}_j, \underline{z}_j) \oplus \underline{z}_j G^*_A] G_A) P_A = \underline{z}'_j \oplus f(\underline{m}_j, \underline{z}_j) E'_A, \tag{9}$$

where $\underline{z}'_j = \underline{z}_j P_A \oplus \underline{z}_j G^*_A G_A P_A$ and $E'_A = G_A P_A$.

Thus if E_A (respectively E'_A) and at least one $\underline{e}'_j$ (respectively $\underline{z}'_j$) are known, then the user's signature can be universally forged in Xinmei's scheme (respectively the other scheme). For example, if the message $\underline{m}_l$ is to be signed, then the cryptanalyst can produce the signature $\underline{c}_l$ as $\underline{c}_l = \underline{e}'_j \oplus \underline{m}_l E_A$ in Xinmei's scheme (respectively as $\underline{c}_l = \underline{z}'_j \oplus f(\underline{m}_l, \underline{z}_j) E'_A$ in the other scheme). Furthermore, if E_A (respectively E'_A) is known, then $\underline{e}'_j$ (respectively $\underline{z}'_j$) can be readily found. Hence the cryptanalyst needs only to find E_A (respectively E'_A).

Let $\underline{c}_j$ and $\underline{c}_{j'}$ (respectively $\underline{v}_j$ and $\underline{v}_{j'}$) be the signatures of the messages $\underline{m}_j$ and $\underline{m}_{j'}$ under Xinmei's scheme (respectively the other scheme), where $\underline{e}_j$ (respectively $\underline{z}_j$) is the error vector used in both signatures. Then $\underline{c}_j \oplus \underline{c}_{j'} = (\underline{m}_j \oplus \underline{m}_{j'}) E_A$ (respectively $\underline{v}_j \oplus \underline{v}_{j'} = (\underline{m}_j \oplus \underline{m}_{j'}) E'_A$). Now the cryptanalyst needs k pairs of signatures such that each pair uses the same error vector. The k expressions $\{\underline{c}_j \oplus \underline{c}_{j'} = (\underline{m}_j \oplus \underline{m}_{j'}) E_A\}_{1 \le j, j' \le k}$ (respectively $\{\underline{v}_j \oplus \underline{v}_{j'} = (\underline{m}_j \oplus \underline{m}_{j'}) E'_A\}_{1 \le j, j' \le k}$) form a linear system which allows us to solve for E_A (respectively E'_A) in $O(k^3)$ provided that set of the messages $\{\underline{m}_j \oplus \underline{m}_{j'}\}_{1 \le j, j' \le k}$ are linearly independent. The cryptanalyst can then find one or more $\underline{e}'_j$'s (respectively $\underline{z}'_j$'s) by using E_A (respectively E'_A).

The efficiency of this attack can be expressed as the number of signatures l that must be obtained before the attack succeeds. It is assumed that the signatures generated by a user are uniformly distributed. Let N be the number of possible error vectors that can be invoked by the signer. Clearly

$$N = \binom{n}{t'_A}. \tag{10}$$

The problem exhibits a great resemblance to the birthday paradox. We expect the number of signatures required for this universal forgery attack to be $O(\sqrt{N})$. To support this argument, we take an approach similar to the one given in [4, pp. 279–281].

Let the signatures be grouped into r sets such that each set contains s signatures, where $s^2 \le N$. For any two sets, the total number of comparisons is s^2 and the probability that a comparison would yield a match is $1/N$ (the match event refers to the event when two signatures from two different sets have the same error vector, a match within a set is not considered here). Thus the probability of a match between any two sets is approximately s^2/N. By making $s = \sqrt{N}$, there is then a match between any two sets with overwhelming probability (it was mentioned in [9] that there are at least three elements in common between any two such sets). Furthermore, r is chosen such that $\binom{r}{2} \ge k$, and thus $r = \left\lceil \frac{1+\sqrt{1+8k}}{2} \right\rceil$. The total number of signatures needed is thus $l = O(r\sqrt{N})$.

The attack requires two tables: one containing the signatures and another containing the error vectors. The space requirement is thus $O[l(n + \lceil log_2 N \rceil)]$ bits, where each error vector requires $\lceil log_2 N \rceil$ bits. $O(Nk)$ comparisons are needed for this attack. But this complexity can be dramatically reduced, however, by sorting the whole l signatures instead of comparing the elements one by one. This sorting technique would have a complexity of $O(l \log_2 l)$ comparisons and since each comparison involves two $\lceil log_2 N \rceil$-bit numbers. It follows that the bit complexity is $O(\lceil log_2 N \rceil l \log_2 l)$ bit operations.

Finally we must consider the question of whether the set of the messages $\{\underline{m}_j \oplus \underline{m}_{j'}\}_{1 \le j, j' \le k}$ are linearly independent. It is to be noted that the number of $k \times k$ binary invertible matrices is $0.29 \times 2^{k^2}$, and the number of $k \times k$ binary matrices is 2^{k^2}. Thus the probability of randomly selecting any $k \times k$ binary matrix and having it be invertible is thus 0.29, and the expected number of repetitions is thus 3.4. The number of signatures that must be collected is thus increased on the average by a factor slightly more than 3.

4.2 Attack II

In Xinmei's scheme, $\underline{c}_j$ must satisfy the equations $\underline{c}_j T_A = \underline{e}_j H_A^T$ and $\underline{c}_j J_A = \underline{e}_j W_A \oplus \underline{m}_j$. This leads to $\underline{c}_j [T_A \mid J_A] = [\underline{e}_j H_A^T \mid \underline{e}_j W_A \oplus \underline{m}_j]$, or simply $\underline{c}_j X = Y$, where $Y = [\underline{e}_j H_A^T \mid \underline{e}_j W_A \oplus \underline{m}_j]$ is an $n \times n$ matrix and $X = [T_A \mid J_A]$ is an $n \times n$ matrix which is publicly known. Similarly, in the authors' scheme, $\underline{v}_j$ should satisfy $\underline{v}_j H'_A = \underline{z}_j H_A^T$ and $\underline{v}_j G'_A = f(\underline{m}_j, \underline{z}_j)$. This leads to $\underline{v}_j [H'_A \mid G'_A] = [\underline{z}_j H_A \mid f(\underline{m}_j, \underline{z}_j)]$, or simply $\underline{v}_j X' = Y'$, where $Y' = [\underline{z}_j H_A \mid f(\underline{m}_j, \underline{z}_j)]$ is an $n \times n$ matrix and $X' = [H'_A \mid G'_A]$ is an $n \times n$ matrix which is publicly known.

The analyst computes Y (respectively Y'), where $\underline{m}_j$ is the message to be forged and $\underline{e}_j$ (respectively $\underline{z}_j$) has weight t'_A, then $\underline{c}_j$ (respectively $\underline{v}_j$) can be easily obtained as $\underline{c}_j = YX^{-1}$ (respectively $\underline{v}_j = Y'X'^{-1}$), provided that X (respectively X') is full rank. The following lemma shows that X and X' are both full rank matrices, and thus signatures can be universally forged in both schemes.

Lemma 1 *X and X' are full rank matrices.*

Proof: We will only prove the lemma for X; the result for X' can be proven in a similar manner. Let $\underline{w}$ be an n-bit column vector and partition $\underline{w}$ into two vectors, an $(n-k)$-bit column vector $\underline{w}_1$ and a k-bit column vector $\underline{w}_2$. Thus $X\underline{w} = [T_A \mid J_A]\underline{w} = [P_A^{-1} H_A^T \mid P_A^{-1} G_A^* S_A^{-1}]\underline{w} = P_A^{-1} H_A^T \underline{w}_1 \oplus P_A^{-1} G_A^* S_A^{-1} \underline{w}_2$. If it holds that $X\underline{w} = 0_{n \times 1}$ (the all-zero n-bit column vector) if and only if $\underline{w} = 0_{n \times 1}$, then X has rank n. For $\underline{w} = 0_{n \times 1}$, we have $X\underline{w} = 0_{n \times 1}$. For $X\underline{w} = 0_{n \times 1}$, we have $P_A^{-1} H_A^T \underline{w}_1 \oplus P_A^{-1} G_A^* S_A^{-1} \underline{w}_2 = 0_{n \times 1}$. Premultiplying by $G_A P_A$, we obtain $G_A H_A^T \underline{w}_1 \oplus G_A G_A^* S_A^{-1} \underline{w}_2 = 0_{k \times 1}$. Since H_A is the null space of G_A, we have $G_A H_A^T = 0_{k \times (n-k)}$ and hence $G_A G_A^* S_A^{-1} \underline{w}_2 = 0_{k \times 1}$, or $S_A^{-1} \underline{w}_2 = 0_{k \times 1}$ and thus $\underline{w}_2 = 0_{k \times 1}$, for S_A is nonsingular. Hence $H_A^T \underline{w}_1 = 0_{n \times 1}$. Since H_A^T has rank $n-k$, then $\underline{w}_1 = 0_{(n-k) \times 1}$. ∎

Li [7] mentioned that if the public key in Xinmei's scheme is chosen such that X is not a full rank matrix, then the scheme is secure. Lemma 1, on the contrary, shows that X is always full rank, regardless of the selection of the public key.

5 Conclusion

Examining the previous digital signature schemes based on linear error-correcting block codes, the following can be concluded:

- The linearity of the code allows selective forgery [5]. This, however, can be prevented by signing the image of the message under a nonlinear noninvertible transformation [5] instead of signing the message itself. It is essential that the transformation is a function of the error vector [3] to prevent the chosen-message attack devised in [1].
- Revealing the error vectors permits universal forgery as shown in this paper (attack I). This attack can be prevented if the error vectors are not revealed or the parameters of the code are chosen properly to make this attack infeasible.
- Revealing information about the right inverse of the generator matrix is equivalent to revealing some information about k linearly independent columns of the generator matrix. This is the reason for the success of the direct attack on the Xinmei and Harn-Wang schemes [12] and the universal forgery attack (attack II) described here. (It is to be noted that the probabilistic ciphertext-only attacks launched on McEliece's system [8] as described in [6, 8, 11] are all based on searching for k linearly independent columns of the generator matrix). Thus the public key should not contain information (even in scrambled form) about k linearly independent columns of the generator matrix.

References

1. M. Alabbadi and S. B. Wicker. Cryptanalysis of the Harn and Wang modification of the Xinmei digital signature scheme. *Electronics Letters*, 28(18):1756–1758, 27th August 1992.
2. M. Alabbadi and S. B. Wicker. Security of Xinmei's digital signature scheme. *Electronics Letters*, 28(9):890–891, 23rd April 1992.
3. M. Alabbadi and S. B. Wicker. Digital signature schemes based on error-correcting codes. In *IEEE International Symposium on Information Theory*, January 17-22 1993. San Antonio, Texas, U.S.A.
4. D. W. Davies and W. L. Price. *Security for Computer Networks*. John Wiley and Sons, 1989.
5. L. Harn and D. -C. Wang. Cryptanalysis and modification of digital signature scheme based on error-correcting codes. *Electronics Letters*, 28(2):157–159, 16th January 1992.
6. P. J. Lee and E. F. Brickell. An obsevation on the security of McEliece's public-key cryptosystem. In C. G. Gunther, editor, *Lecture Notes in Computer Science # 330, Advances in Cryptology-Eurocrypt '88 Proceedings*, pages 275–280, Davos, Switzerland, May 25-27 1988. Springer-Verlag.

7. Yuan-Xing Li. An attack on Xinmei's digital signature scheme. In *IEEE International Symposium on Information Theory*, January 17-22 1993. San Antonio, Texas, U.S.A.
8. R. J. McEliece. Public-key cryptosystem based on algebraic coding theory. JPL DSN Progress Report 42-44, Jet Propulsion Laboratory, California Institute of Technology, Pasadena, CA, U.S.A, Jan. & Feb. 1978. Pages 114-116.
9. J. Meijers and J. van Tilburg. On the Rao-Nam private-key cryptosystem using linear codes. In *IEEE International Symposium on Information Theory*, page 126, June 24-28 1991. Budapest, Hungary.
10. National Bureau of Standard. *Data Encryption Standard, Federal Information Processing Standard (FIPS) Publication 46*, January 1977. U.S. Department of Commerce, Washington, D.C.
11. J. van Tilburg. On the McEliece public-key cryptosystem. In S. Goldwasser, editor, *Lecture Notes in Computer Science # 403, Advances in Cryptology-Crypto '88 Proceedings*, pages 119–131, Santa Barbara, Ca., Aug. 21-25 1988. Springer-Verlag.
12. J. van Tilburg. Cryptanalysis of Xinmei digital signature scheme. *Electronics Letters*, 28(20):1935–1936, 24th September 1992.
13. W. Xinmei. Digital signature scheme based on error-correcting codes. *Electronics Letters*, 26(13):898–899, 21st June 1990.

On Message Protection in Cryptosystems Modelled as the Generalized Wire-Tap Channel II

Miodrag J. Mihaljević*,**

**Institute of Applied Mathematics and Electronics,
Institute of Mathematics, Academy of Arts and Sciences,
Belgrade, Yugoslavia**

Abstract. A novel approach to the analysis of message security in cryptosystems which can be modelled as the generalized wire-tap channel II is considered. Roughly speaking, the codewords transmitted through this channel are degraded by the deletion of a number of bits and by the complementing of a number of the remaining bits. It is assumed that the channel output codewords are available to a cryptanalyst. The security of the messages is measured by the uncertainty after certain minimum distance decoding (MDD) procedures are applied to the received codewords. A novel distance measure relevant to MDD is proposed, and the remaining uncertainty after MDD is analyzed. An extension of the approach is also considered.

1 Introduction

According to [1, pp. 501-502], cryptosystems may be classified into the following three major types: Conventional Cryptosystems, Public-Key Cryptosystems and the Wire-Tap Channel. A nonasymptotic analysis of the information protection at the output of the wire-tap channel, introduced in [2], is presented in [3]. Two criteria for information protection are discussed in [3]. The first concerns the channel capacity for sources with given redundancy. The second concerns the probability that the real transmitted message assumes a value in the list, of given size, which was formed during optimal list decoding.

The wire-tap channel II was introduced and analyzed in [4]. In the binary version, the sender has k bits of information to convey to the receiver by n uses of the channel. The adversary can listen to any s bits of his choosing. The transmission channel is assumed to be noiseless, so correct decoding is not the problem here. The focus of the problem is to prevent the adversary from gaining too much information. Accordingly, at the output of the wire-tap channel, for any n-bit input vector the channel output is an s-bit vector obtained by excluding

* This research was supported by the Science Fund, grant #0403, through Institute of Mathematics, Academy of Arts and Sciences, Belgrade.
** Mailing Address: Solina 4, 11040 Belgrade, Yugoslavia

$n-s$ bits on known positions from the input vector. One of the schemes from [4] uses an $(n, n-k)$ binary linear code C. The code has 2^k cosets, each representing a binary k-typle. If the sender wants to transmit k bits of information to the receiver, he selects a random vector in the corresponding coset. The legitimate receiver can determine the corresponding coset of the received vector because its channel is noiseless. The adversary has full knowledge of the code C, but no access to the random process that selects a vector from the coset.

Recently, a number of papers related to the wire-tap channel II have appeared. In [5] a generalization of the minimum distance of a binary code is introduced. For any code D, let $X(D)$, the support size of D, be the number of positions where not all the codewords of D are zero. For an (n, k) code C and any r, where $1 \leq r \leq k$, the rth generalized Hamming weight is defined by $d_r(C) = min\{X(D) : D \textit{ is an } (n,r) \textit{ subcode of } C\}$. In particular, the minimum distance of C is $d_1(C)$, and the weight hierarchy of C is the set $d_r(C) : 1 \leq r \leq k$. It was shown in [5] that for the previously described special scheme the following is valid: If the adversary is allowed to tap s bits (of his choice) from the transmitter he will obtain r bits of information, if and only if $s \geq d_r(C)$.

Therefore, in order to analyze the performance of C on the wire-tap channel of Type II, assuming the considered encoding scheme, it is necessary to find the weight hierarchy of C or at least to find good bounds on the generalized Hamming weights. These problems are considered in [6] and [7].

This paper is devoted to the wire-tap channel II area but its objective and the employed approach are different in comparison with the previously discussed results.

In this paper a more complex type of wire-tap channel II is introduced and the general encoding of k information bits into an n-dimensional binary vector is assumed. This scheme is analyzed from the cryptanalyst point of view, employing a novel approach based on the sequence comparison concept. The protection of messages against a wire-tapper (cryptanalyst) is considered in the cryptosystems which can be modelled as a so called generalized wire-tap channel II. In this channel, for any n-bit input vector the corresponding output is an s-bit vector $(s < n)$ obtained by excluding $n-s$ bits on unknown positions from the input vector and complementing a number of the remained s bits (also on unknown positions). The protection of the messages is measured by the wire-tapper's uncertainty after a minimum distance decoding (MDD) procedure.

In Section 2, the generalized wire-tap channel II and the wire-tapper's minimum distance decoding procedure are presented. A novel distance measure relevant to the MDD is proposed in Section 3, and the uncertainty after MDD is analyzed in Section 4. Illustrative numerical examples are presented in Section 5. A modification of the channel, where an extra constraint is involved, and corresponding results are given in Section 6. The conclusions are stated in Section 7.

2 Preliminaries

In this section, a generalized wire-tap channel II is specified. A decoding procedure for the codewords received through this channel and the problem of message security are then stated.

2.1 Generalized Wire-Tap Channel II

For the purpose of further consideration, a basis for the sequence comparison approach will be given. Note that it is often necessary to compare two or more sequences and to measure the extent to which they differ. Sequence comparison deals with the comparisons that arise when the correspondence between the elements is not known in advance (perhaps because some underlying correspondence has been disturbed by the loss of elements in one or both sequences). The general approach used in sequence comparison is to seek the appropriate correspondence by optimizing over all possible correspondences that satisfy suitable conditions, such as preserving the order of the elements in the sequence. In order to measure the similarity/dissimilarity (between the sequences) in the sequence comparison, it is assumed that one sequence can be obtained from the other by a number of edit-operations. The standard edit-operations that transform one sequence into another are symbol substitution, deletion and insertion. A review of the sequence comparison techniques and applications is presented in [8]. According to [8], one of the most widely used distances is the Levenshtein Distance (LD) [9]. The LD between two sequences is defined as the minimum number of standard edit-operations (substitution, deletion and insertion) required to transform one sequence into another.

The generalized wire-tap channel II will be specified in terms of the sequence comparison. We assume that the wire-tapper observes the codewords through the following channel.

Definition 1. Generalized wire-tap channel II is a channel which modifies an n-bit input codeword into an s-bit channel output, $n > s$, in a such way that the channel output can be transformed into the corresponding input codeword by the following edit operations: $\Delta = n - s$ insertions, and no more than η complementations.

2.2 MDD and Message Protection

In this paper, a general encoding of k information bits into an n-dimensional binary vector is under consideration, so that there are 2^k distinct codewords.

After a codeword is received through the generalized wire-tap channel II, we assume that the wire-tapper applies a standard minimum distance decoding (MDD) procedure because, for the time being, no more efficient decoding procedure is known for a general binary block code. The MDD means that the received codeword has to be compared with all possible codewords using an appropriate distance measure, and after all possible trials, the codeword at the minimum distance from the received word has to be accepted as the codeword that was sent. Related to the MDD for the generalized wire-tap channel II, note that construction of the appropriate distance measure is the main problem which has to be solved.

When more than one codeword has the same minimum distance to the received codeword, uncertainty exists about the transmitted codeword. So, a natural way for measuring message protection (or security) is in the wire-tapper's uncertainty after the MDD procedure.

3 A Novel Distance Measure for MDD

In this section a novel distance measure relevant to the wire-tapper's MDD procedure is proposed. Here, the word distance is not restricted only to indicate a function which satisfies the metric conditions. The distance measure is defined so that it enables the statistical distinction between the following two cases: first when the received codeword is compared with the corresponding transmitted codeword, and second when the received codeword is compared with any other codeword. According to the generalized wire-tap channel II definition, the distance has to reflect the fact that the received codeword could be transformed into the transmitted one using a given number of insertions and no more than a given number of substitutions. The distance measure defined here originates from the LD calculation procedure proposed and discussed in [10] - [11], and from the results given in [12].

Denote by $LD(\{a_i\}_{i=0}^{n}, \{b_i\}_{i=0}^{s})$ the LD between two arbitrary binary sequences $\{a_i\}_{i=0}^{n}$ and $\{b_i\}_{i=0}^{s}$, $n > s$. Assume that $\Delta \leq n - s$ and η are given positive integers such that $LD(\{a_i\}_{i=0}^{n}, \{b_i\}_{i=0}^{s}) \geq \Delta + \eta$. Then, a novel distance measure $d = d(\{a_i\}_{i=0}^{n}, \{b_i\}_{i=0}^{s})$ relevant to the MDD is defined by the following.

Definition 2.

$$d = s - max\{\ell \mid LD(\{a_i\}_{i=0}^{\Delta+\ell}, \{b_i\}_{i=0}^{\ell}) = \Delta + \eta\} \; . \tag{1}$$

So the defined distance measure d , according to the Definition 1, provides zero-distance between the received and transmitted codewords.

Note that d can be efficiently calculated using the procedure for the Levenshtein distance calculation presented in [10]. In the binary case when only insertions (or deletions) and substitutions of the symbols are allowed, the main part of the algorithm [10], when the sequences lengths are n and s , is the generation of an $(n - s) \times s$-dimensional matrix $u = [u(i,j)]_{i=0}^{n-s}{}_{j=0}^{s}$ such that:

$$\begin{aligned}
&u(i,0) = i, && i = 0,1,\cdots,n-s,\\
&u(0,j) = u(0,j-1) + (a_j \oplus b_j), && j = 1,2,\cdots,s,\\
&u(i,j) = min\{[u(i-1,j)+1], && \\
&\qquad [u(i,j-1) + (a_{i+j} \oplus b_j)]\}, && i = 1,2,\cdots,n-s,\\
& && j = 1,2,\cdots,s.
\end{aligned} \tag{2}$$

It can be shown that

$$d = s - max\{\ell \mid u(\Delta,\ell) - u(\Delta,0) = \eta\}, \tag{3}$$

Note that d could be considered as a realization of an integer stochastic variable D . Regarding the analytical treatment of the relevant probability distributions related to the standard distance measures for sequence comparison it is noted on p.352 of [8] that the derivation of exact mathematical results seems difficult and many interesting questions remain unanswered. On the other hand, in the next section it will be shown that a significance aspect of the distance measure given by Definition 2, in comparison with the standard distance measures, is that the new one enables analytical consideration of an important characteristic of the relevant probability distributions.

4 On Message Uncertainty After MDD

Recall that, according to Definitions 1 and 2, we have zero-distance ($d = 0$) between the transmitted and received codewords. On the other hand, the wire-tapper is confused when more than one codeword is at zero-distance from the received codeword. Note that the number No of codewords at zero-distance from a received codeword is a realization of a stochastic variable.

So, in order to estimate the wire-tapper's uncertainty we have to estimate, for given parameters k, n, Δ and η, the expected number of codewords at zero-distance from the received codeword. This problem could be solved numerically, employing a histogram estimation approach (and the channel model according to Definition 1), for example, but this approach would be time consuming for some values of the parameters k, n, Δ and η. For the further analytical consideration of the wire-tapper's uncertainty we use the following assumption.

Assumption 1. When $n >> k$, the encoding scheme and the noise on the wire-tapper's channel provide that the comparison between the received and any other codeword except the transmitted one is equivalent to the comparison of two random, mutually independent, binary balanced sequences.

An estimation of a lower bound on the wire-tapper's uncertainty is given by the following theorem.

Theorem 1. *After the MDD procedure, based on the distance d, when Assumption 1 holds and* $\Delta >> \eta$, *a lower bound on the expected number* $\bar{No}$ *of the codewords at zero-distance from the received codeword can be estimated by the following:*

$$\bar{No} \geq 1 + \left\lfloor \frac{(2^k - 1) \sum_{i=0}^{\Delta^*} 2^{-(s+i)} \binom{s+i}{i}}{\left[\sum_{i=0}^{s-1} 2^{-(\Delta^*+i)} \binom{\Delta^*+i}{\Delta^*}\right] + \left[\sum_{i=0}^{\Delta^*} 2^{-(s+i)} \binom{s+i}{i}\right]} \right\rfloor, \tag{4}$$

where

$$\Delta^* = \Delta + \lfloor \eta(1 + \log_2 n - \log_2 \eta) \rfloor \ , \ \eta > 0 \ , \tag{5}$$

$\Delta^* = \Delta$ *for* $\eta = 0$, $s = n - \Delta$, *and* $\lfloor \cdot \rfloor$ *denotes the integer part.*

A sketch of the theorem proof is given in the appendix.

5 Numerical Examples

In this section a number of numerical examples related to the wire-tapper's uncertainty are given. An objective of this consideration is to present quantitative relationships between the wire-tapper's uncertainty and the estimation of its lower bound based on Theorem 1. For different values of parameters k, n, Δ, and η, the MDD procedures based on the distance d were realized and the numbers of codewords at zero-distance to a received one were counted. On the other hand, the corresponding lower bound estimations were calculated according to Theorem 1.

The numerical examples show the average number of codewords at zero-distance from a received one, after the MDD procedure, and the lower bound estimate of its expected value calculated according to Theorem 1 as functions of the parameters k, n, Δ, and η. The results are summarized in the following tables.

The first three tables present an extract of the average number of codewords at zero-distance from a received one, counted on a sample of the MDD procedures and the estimation of its lower bound, calculated according to the Theorem 1, as functions of the allowed number of complementations when other parameters are fixed.

Table 1. The average number of codewords at zero-distance from a received one, counted on a sample and estimation of its lower bound calculated according to Theorem 1 as a function of η when $n = 64$, $\Delta = 24$, and $k = 12$.

η	*average counted number of codewords at zero − distance*	*estimation of lower bound on expected number of codewords at zero − distance*
0	117	97
1	639	494
2	1590	1208
3	2631	2144

Table 2. The average number of codewords at zero-distance from a received one, counted on a sample and estimation of its lower bound calculated according to Theorem 1 as a function of η when $n = 48$, $\Delta = 16$, and $k = 12$.

η	*average counted number of codewords at zero − distance*	*estimation of lower bound on expected number of codewords at zero − distance*
0	51	45
1	439	371
2	1126	1110

Table 3. The average number of codewords at zero-distance from a received one, counted on a sample and estimation of its lower bound calculated according to Theorem 1 as a function of η when $n = 192$, $\Delta = 64$, and $k = 12$.

η	*average counted number of codewords at zero − distance*	*estimation of lower bound on expected number of codewords at zero − distance*
0	1	1
1	1	1
2	3	2
3	10	8
4	29	21
5	141	57
6	335	133
7	706	240
8	1196	402
9	1855	631
10	2510	927
11	3121	1283
12	3502	1683

The next two tables present typical examples of the average number of codewords at zero-distance from a received codeword counted on a sample of the MDD procedures and estimation of its lower bound calculated according to Theorem 1 as a functions of number of deleted bits in the generalized wire-tap channel II and the codeword dimension, respectively, when other parameters are fixed.

Table 4. The average number of codewords at zero-distance from a received one counted on a sample and estimation of its lower bound calculated according to Theorem 1 as a function of Δ when $n = 256$, $\eta = 5$, and $k = 12$.

Δ	*average counted number of codewords at zero − distance*	*estimation of lower bound on expected number of codewords at zero − distance*
80	1	1
85	9	4
90	85	21
95	446	94
100	1228	327
105	2486	856
110	3452	1707

Table 5. The average number of codewords at zero-distance from a received one counted on a sample and estimation of its lower bound calculated according to Theorem 1 as a function of n when $\Delta = 64$, $\eta = 5$, and $k = 12$.

n	*average counted number of codewords at zero − distance*	*estimation of lower bound on expected number of codewords at zero − distance*
160	3126	1746
176	1121	385
192	141	57
208	5	3
224	1	1

Note that the lower bound estimation given by Theorem 1 could be considered as a product of 2^k and a certain function of n, Δ, η. So, the lower bound can be easily estimated, for example in the cases presented in Table 1 - Table 5 when instead of $k = 12$ we have $k >> 12$, which is an extremely time consuming case for the sample consideration.

6 On a Constrained Case

In this section, an extension of the presentations in Sections 2-4 is given. We shall consider a modification of the channel specified by Definition 1. The difference is in the following: Instead of Δ unconstrained insertions we assume that Δ insertions are allowed, but under the constraint that consecutive insertions are forbidden. Accordingly, note the following background. The Constrained Levenshtein Distance (CLD) is also considered in the literature (see [13] and [14], for example). The CLD is equal to the LD where the number and/or positions of the edit operations satisfy certain conditions. Here, the CLD is defined as the minimum number of insertions subject to the constraint that successive insertions are not allowed, and complementations required to obtain one (longer) binary sequence from the other (shorter one). The efficient algorithm for the CLD calculation is proposed in [14].

Denote by $CLD(\{a_i\}_{i=0}^{n}, \{b_i\}_{i=0}^{s})$ the CLD between two arbitrary binary sequences $\{a_i\}_{i=0}^{n}$ and $\{b_i\}_{i=0}^{s}$, $n > s$. Assume that $\Delta \leq n - s$ and η are positive integers such that $CLD(\{a_i\}_{i=0}^{n}, \{b_i\}_{i=0}^{s}) \geq \Delta + \eta$. Then, the constrained distance measure $d^* = d^*(\{a_i\}_{i=0}^{n}, \{b_i\}_{i=0}^{s})$, relevant to the MDD in the case considered here, is defined by the following.

Definition 3.

$$d^* = s - max\{\ell \mid CLD(\{a_i\}_{i=0}^{\Delta+\ell}, \{b_i\}_{i=0}^{\ell}) = \Delta + \eta\} \ . \tag{6}$$

According to [14], it could be shown that an efficient way for d^* to be calculated is given by:

$$d^* = s - max\{\ell \mid u_c(\Delta, \ell) - u_c(\Delta, 0) = \eta\}, \tag{7}$$

where $u_c(i,j)$, $i = 0, 1, \cdots, n-s$, $j = 0, 1, \cdots, s$, are defined by the following:

$$\begin{aligned} u_c(i,0) &= i, & i &= 0,1,\cdots,n-s, \\ u_c(0,j) &= u_c(0,j-1) + (a_j \oplus b_j), & j &= 1,2,\cdots,s, \\ u_c(i,j) &= min\{[u_c(i-1,j-1) + (a_{i+j-1} \oplus b_j) + 1], & & \\ & \qquad [u_c(i,j-1) + (a_{i+j} \oplus b_j)]\}, & i &= 1,2,\cdots,n-s, \\ & & j &= 1,2,\cdots,s. \end{aligned} \tag{8}$$

The wire-tapper's uncertainty and the message protection, in this case, could be estimated in a similar way as it was noted in Section 4. Based on the considered channel model and Definition 3, the expected number of codewords at zero-distance from the received codeword, which determines the uncertainty, could be estimated by employing the standard histogram approach. For the time being, no one relevant analytical result is known, and this is one of the open problems.

7 Conclusions

In this paper, message security in cryptosystems which can be modelled as the generalized wire-tap channel II is considered. In this channel, for any n-bit input vector the corresponding output is an s-bit vector ($s < n$) obtained by excluding $n - s$ bits at unknown positions from the input vector and complementing a number of the remaining s bits, also at unknown positions. The general encoding of k information bits into the n-bit codeword is assumed. The problem of message protection is considered by employing a novel approach based on the sequence comparison concept.

Message security is measured by the wire-tapper's uncertainty after certain minimum distance decoding (MDD) procedure is applied to the received codewords. A novel distance measure relevant for the MDD is proposed, and an efficient procedure for its calculation is given. The uncertainty remaining after the MDD is analyzed theoretically and experimentally. An estimation of a lower bound on the expected number of codewords at zero-distance from a received codeword is derived. The MDD procedures based on the novel distance and calculations of the lower bound uncertainty estimations are illustrated through a number of numerical examples. A modification of the channel model, related to the case when an extra constraint on the edit-operations is imposed, is also considered, and a relevant distance measure, together with an efficient procedure for its calculation, are presented.

8 Appendix

Sketch of the Theorem 1 proof. First consider a situation when only Δ insertions are allowed for the transformation of a sequence $\{b_i\}_{i=1}^{\ell}$ into an longer sequence $\{a_i\}_{i=1}^{\Delta+\ell}$, assuming that ℓ has maximum possible value. Suppose that for some index i values $i1$ and $i2$, $i1 > i2$, after the partial transformation of $\{b_i\}_{i=1}^{i2}$ into $\{a_i\}_{i=1}^{i1}$, we have one of the following two situations:

$$a_{i1+1} = 0, \qquad b_{i2+1} = 1, \qquad a_{i1+i} = b_{i2+i} = 0, \qquad i = 2, 3, ..., \delta,$$

or

$$a_{i1+1} = 1, \qquad b_{i2+1} = 0, \qquad a_{i1+i} = b_{i2+i} = 1, \qquad i = 2, 3, ..., \delta,$$

where δ is a parameter.

Note that any of these situations produces a subsequence of at least δ insertions in the edit-sequence which transforms maximum length $\{b_i\}$ into the corresponding $\{a_i\}$. On the other hand, if a complementation is allowed, in the considered situation we have that after the complementation of b_{i2+1} the subsequence of successive insertions in the edit-sequence will be exchanged with a subsequence of δ successive agreements. So, according to the distance d calculation procedure (see (2)-(3)), it can be shown that the expected implication of one complementation is equivalent to increasing the allowed number of insertions for at least 2δ instead of this complementation. Note that 2δ is only a lower bound on the expected value because the distance calculation procedure seeks the appropriate correspondence by optimizing over all possible correspondences, and involving one complementation we increase the number of possibilities. Obviously, an estimation of the the expected value δ, when $\Delta >> 1$ is equal to the greatest integer δ such that:

$$2(2^{-2\delta} n) \geq 1\,, \tag{9}$$

so that an estimation of the expected effect on the distance d value when only one complementation is involved is equivalent to allowing Δ' insertions for the transformation, where

$$\Delta' = \Delta + (1 + \lfloor \log_2 n \rfloor)\,, \tag{10}$$

instead of Δ insertions and one complementation.

Following the given ideas and assuming $\Delta >> \eta$, $\eta > 0$, it can be shown that if η complementations are allowed, a lower bound on the number of possibilities such that $d = 0$ can be estimated assuming that the number of allowed insertions is

$$\Delta^* = \Delta + \lfloor \eta(1 + \log_2 n - \log_2 \eta) \rfloor\,, \tag{11}$$

instead of Δ insertions and η complementations.

On the other hand, according to [12, Theorem 1], the probability that $d = 0$ when the sequences under comparison are uncorrelated, the shorter sequence length is equal to s, and Δ^* insertions are allowed is given by the following.

$$Pr(D=0) = \frac{\sum_{i=0}^{\Delta^*} 2^{-(s+i)} \binom{s+i}{i}}{\left[\sum_{i=0}^{s-1} 2^{-(\Delta^*+i)} \binom{\Delta^*+i}{\Delta^*}\right] + \left[\sum_{i=0}^{\Delta^*} 2^{-(s+i)} \binom{s+i}{i}\right]} \tag{12}$$

Finally, the theorem statement is a direct consequence of (12) and the fact that (11) yields an estimate of the lower bound on the expected number of insertions which are required to ensure the same effect as Δ insertions and η complementations.

Q.E.D.

References

1. R.E.Blahut, *Digital Transmission of Information.* Addison-Wesley, Reading MA, 1990.
2. A.D.Wyner, "The wire-tap channel", *Bell Syst. Tech. Journal*, vol. 54, pp. 1355-1387, 1975.
3. V.Korzhik and V.Yakovlev, "Nonasymptotic estimates of information protection efficiency for the wire-tap channel concept", *Advances in Cryptology - AUSCRYPT "92, Lecture Notes in Computer Science*, vol. 718, pp. 185-195, 1993.
4. L.H.Ozarow and A.D.Wyner, "Wire-tap channel II", *AT & T Bell Lab. Tech. Journal*, vol. 63, pp. 2135-2157, Dec. 1984.
5. V.K.Wei, "Generalized Hamming weights for linear codes", *IEEE Trans. Inform. Theory*, vol. 37, pp. 1412-1418, Sep. 1991.
6. G.L.Feng, K.K.Tzeng and W.K.Wei, "On the generalized Hamming weights of several classes of cyclic codes", *IEEE Trans. Inform. Theory*, vol. 38, pp. 1125-1130, May 1992.
7. T.Helleseth, T.Kløve, O.Ytrehys, "Generalized Hamming weights of linear codes", *IEEE Trans. Inform. Theory*, vol. 38, pp. 1133-1140, May 1992.
8. D.Sankoff and J.B.Kruskal, *Time Warps, String edits and macromolecules: The theory and practice of sequence comparison.* Reading, MA: Addison-Wesley, 1983.
9. A.Levenshtein, "Binary codes capable of correcting deletions, insertions, and reversals", *Sov. Phy. Dokl.*, vol. 10, pp. 707-710, 1966.
10. B.J.Oommen, "Recognition of noisy subsequences using constrained edit distance", *IEEE Trans. Pattern Analysis Mach. Intell.*, vol. PAMI-9, pp. 676-685, Sep. 1987.
11. B.J.Oommen, "Correction to Recognition of noisy subsequences using constrained edit distance", *IEEE Trans. Pattern Analysis Mach. Intell.*, vol. PAMI-10, pp. 983-984, Nov. 1988.
12. M.Mihaljević, "An approach to the initial state reconstruction of a clock-controlled shift register based on a novel distance measure", *Advances in Cryptology - AUSCRYPT "92, Lecture Notes in Computer Science*, vol. 718, pp. 349-356, 1993.
13. B.J.Oommen, "Constrained string editing", *Inform.Sci.*, vol.40, pp.267-284, 1986.
14. J.Golić and M.Mihaljević, "A generalized correlation attack on a class of stream ciphers based on the Levenshtein distance", *Journal of Cryptology*, vol. 3 (3), pp.201-212, 1991.

How to avoid the Sidel'nikov-Shestakov attack

Ernst M. Gabidulin[1] and Olaf Kjelsen[2]

[1] Moscow Institute of Physics and Technology
Institutskii per. 9
141700 Dolgoprudnyi RUSSIA
e-mail: gab@re.mipt.su

[2] Swiss Federal Institute of Technology Zurich
Institute for Information Processing
ETH-Zentrum, ETZ F 89, CH-8092 Zurich, Gloriastrasse 35

Abstract. Recently [2] Sidel'nikov and Shestakov showed how to break the Niederreiter Public-Key Cryptosystem. It is proposed to make breaking this PKC based on Generalized Reed-Solomon codes more difficult by randomizing the parity check matrix. The modified Niederreiter system appears to be secure.

1 Introduction

A *Public-Key Cryptosystem* (PKC) based on Linear Generalized Reed-Solomon codes was proposed by Niederreiter [1]. This is a Knapsack-Type Cryptosystem, and there is thus little hope that this PKC is secure. In fact, as with many other knapsack-type PKC's, it was shown by Sidel'nikov and Shestakov [2, 3] that the *Niederreiter* PKC and some of its modification are insecure.

A modification of this system is proposed in this paper. The modified *Niederreiter* PKC described here appears secure, and is, in particular, resistant to the Sidel'nikov-Shestakov attack.

The main idea is as follows. A legal party chooses as an extra secret key some random matrix and adds it to the original parity check matrix to produce a new modified public key. Any visible structure of the public key is thus hidden. Of course, there are strong restrictions when we choose the random matrix, but this is easily handled in practical applications.

The paper is organized as follows. In Section 2, some properties of Generalized Reed-Solomon codes are described. In Section 3, the *Niederreiter* PKC is presented. The Sidel'nikov-Shestakov attack is discussed in Section 4. Our version of the attack differs slightly from the original one. The modification of the *Niederreiter* PKC is given in Section 5.

2 Some Properties of the Generalized Reed-Solomon Codes

Generalized Reed-Solomon (GRS) codes are MDS codes based on Vandermonde-type matrices [4].

Let $GF(q)$ be a finite field. Let α_i, $i = 1, \ldots, n$, be different elements from $GF(q) \cup \{\infty\}$. Let $z_i, i = 1, \ldots, n$, be nonzero, not necessarily different elements from $GF(q)$. The standard parity check matrix for GRS codes has the following form

$$\mathbf{H} = \begin{bmatrix} z_1 & z_2 & \cdots z_n \\ z_1\alpha_1 & z_2\alpha_2 & \cdots z_n\alpha_n \\ z_1\alpha_1^2 & z_2\alpha_2^2 & \cdots z_n\alpha_n^2 \\ \cdots & \cdots & \cdots\cdots \\ z_1\alpha_1^{r-1} & z_2\alpha_2^{r-1} & \cdots z_n\alpha_n^{r-1} \end{bmatrix} . \quad (1)$$

If $\alpha_i = \{\infty\}$ then the ith column of the matrix has all zero entries except the last one, which is equal to z_i. The space of code words consists of all sequences $\overline{g} = (g_1, g_2, \ldots, g_n)$ such that $\overline{g}\mathbf{H}^t = \overline{0}$.

The general parameters for GRS codes are as follows:

- code length $n \leq q + 1$;
- dimension $k = n - r$;
- minimum distance $d = r + 1 = n - k + 1$.
- there exists a Fast Decoding Algorithm for this code (see, for details, [4]).

Of course, a given GRS code has many parity check matrices. Each parity check matrix may be obtained from the standard form in Eq. (1) as a product $\mathbf{H}^* = \mathbf{SH}$ where $\mathbf{S}$ is any $r \times r$ nonsingular matrix. Equivalently, the general parity check matrix $\mathbf{H}^*$ can be written as follows:

$$\mathbf{H}^* = \begin{bmatrix} z_1F_1(\alpha_1) & z_2F_1(\alpha_2) & \cdots z_nF_1(\alpha_n) \\ z_1F_2(\alpha_1) & z_2F_2(\alpha_2) & \cdots z_nF_2(\alpha_n) \\ z_1F_3(\alpha_1) & z_2F_3(\alpha_2) & \cdots z_nF_3(\alpha_n) \\ \cdots & \cdots & \cdots\cdots \\ z_1F_r(\alpha_1) & z_2F_r(\alpha_2) & \cdots z_nF_r(\alpha_n) \end{bmatrix} , \quad (2)$$

where $\{F_1(x), F_2(x), \ldots, F_r(x)\}$ is any set of linearly independent polynomials over the field $GF(q)$ polynomials of degree not greater than $r - 1$.

An important role is played by the so called systematic parity check matrix $\mathbf{H}_{sys}$, which is a matrix of the form

$$\mathbf{H}_{sys} = [\mathbf{E}_r\ \mathbf{R}] =$$

$$\begin{bmatrix} 1\,0\,0 \ldots 0 & R_{1,r+1} & R_{1,r+2} & \ldots & R_{1,n} \\ 0\,1\,0 \ldots 0 & R_{2,r+1} & R_{2,r+2} & \ldots & R_{2,n} \\ 0\,0\,1 \ldots 0 & R_{3,r+1} & R_{3,r+2} & \ldots & R_{3,n} \\ \cdots\cdots & & \cdot & & \cdots \\ 0\,0\,0 \ldots 1 & R_{r,r+1} & R_{r,r+2} & \ldots & R_{r,n} \end{bmatrix} , \quad (3)$$

where $\mathbf{E}_r$ is a $r \times r$ identity matrix and $\mathbf{R}$ is a $r \times (n - r)$ matrix such, that all its square submatrices of any allowable order are non singular.

Lemma 1. *Entries R_{ij} of the systematic parity check matrix in Eq. (3) are equal to*

$$R_{ij} = \frac{z_j}{z_i} \prod_{k=1,k\neq i}^{r} \frac{\alpha_j - \alpha_k}{\alpha_i - \alpha_k}, \ i = 1, 2, \ldots, r; \ j = r+1, r+2, \ldots, n. \tag{4}$$

Proof. It is clear from Eq. (2) that a submatrix of the matrix $\mathbf{H}^*$ consisting of r first columns is non singular. Let $\mathbf{F}^*$ be its inverse. Then $\mathbf{H}_{sys} = \mathbf{F}^*\mathbf{H}^*$. Without any loss of generality we can consider the standard matrix $\mathbf{H}$ (Eq. (1)). To find the matrix $\mathbf{F}$ consider for any $i = 1, 2, \ldots, r$ polynomials of degree $r-1$

$$f_i(x) = \frac{1}{z_i} \prod_{k=1,k\neq i}^{r} \frac{x - \alpha_k}{\alpha_i - \alpha_k} = \sum_{j=0}^{r-1} f_{ij} x^j. \tag{5}$$

Then $\mathbf{F} = [f_{ij}]_{i=\overline{1,r}}^{j=\overline{0,r-1}}$. Calculate the product $\mathbf{FH}$. The ith row of this product is equal to

$$[z_j f_i(\alpha_j), \ j = 1, 2, \ldots, n].$$

It follows from Eq. (5) that $z_j f_i(\alpha_j) = 0$, if $j = 1, \ldots, i-1, i+1, \ldots, r$; $z_j f_i(\alpha_j) = 1$, if $j = i$; $z_j f_i(\alpha_j) = \frac{z_j}{z_i} \prod_{k=1,k\neq i}^{r} \frac{\alpha_j - \alpha_k}{\alpha_i - \alpha_k}$, if $j = r+1, r+2, \ldots, n$. Thus, this is just the ith row of the matrix $\mathbf{H}_{sys}$ in Eq. (3).

Remark. The number of calculations to reduce any matrix $\mathbf{H}^*$ to the systematic form is equal to $O(k^3)$.

Corollary 2.

$$R_{ij} = \frac{u_i v_j}{\alpha_j - \alpha_i}, \ i = 1, 2, \ldots, r; \ j = r+1, r+2, \ldots, n. \tag{6}$$

where

$$\begin{aligned} u_i &= \frac{1}{z_i \prod_{k=1,k\neq i}^{r} (\alpha_i - \alpha_k)}, \ i = 1, 2, \ldots, r; \\ v_j &= z_j \prod_{k=1}^{r} (\alpha_j - \alpha_k), \ j = r+1, r+2, \ldots, n. \end{aligned} \tag{7}$$

Hence, the submatrix $\mathbf{R}$ *of the systematic matrix* $\mathbf{H}_{sys}$ *is the Generalized Cauchy matrix*

$$\mathbf{R} = \mathbf{UCV} = [u_i \delta_{ij}] \left[\frac{1}{\alpha_j - \alpha_i}\right] [v_j \delta_{ij}], \tag{8}$$

where $\mathbf{U}$ *and* $\mathbf{V}$ *are diagonal matrices of order* r *and* $n-r$*, respectively, and* $\mathbf{C}$ *is an* $r \times (n-r)$ *standard rectangular Cauchy matrix. It is well known that all square submatrices of the Cauchy matrix are nonsingular (see, for instance, [4]).*

Remark. It follows from Eq. (7) that

$$\alpha_j = \alpha_1 + \frac{u_1 v_j}{R_{1j}},\ j = r+1, r+2, \ldots, n,$$
$$\alpha_i = \alpha_{r+1} - \frac{u_i v_{r+1}}{R_{i,r+1}},\ i = 2, \ldots, r.$$

Thus the number of calculations to obtain all of the α from the known $\alpha_1, u_i, v_j, R_{ij}$ is equal to $O(n)$.

Corollary 3. *Consider an* $r \times (n-r)$ *matrix* $\mathbf{L} = [L_{ij}] = \left[R_{ij}^{-1}\right] = \left[\frac{\alpha_j - \alpha_i}{u_i v_j}\right]$, $i = 1, 2, \ldots, r$; $j = r+1, r+2, \ldots, n$. *Then all square submatrices of order* 1 *or* 2 *are nonsingular, but any square submatrix of order* 3 *is singular. This means that the matrix* $\mathbf{L}$ *has rank* 2.

Proof. First order submatrices are simply $(\alpha_j - \alpha_i)/u_i v_j \neq 0$ because all α are different. Submatrices of order 2 have the form

$$\begin{bmatrix} \frac{\alpha_j - \alpha_i}{u_i v_j} & \frac{\alpha_k - \alpha_i}{u_i v_k} \\ \frac{\alpha_j - \alpha_m}{u_m v_j} & \frac{\alpha_k - \alpha_m}{u_m v_k} \end{bmatrix}$$

where $r+1 \leq j < k \leq n$, $1 \leq i < m \leq r$. The determinant of this matrix is equal to $(\alpha_m - \alpha_i)(\alpha_k - \alpha_j)/u_i u_m v_j v_k \neq 0$. On the other hand, submatrices of order 3 have the form

$$\begin{bmatrix} \frac{\alpha_j - \alpha_i}{u_i v_j} & \frac{\alpha_k - \alpha_i}{u_i v_k} & \frac{\alpha_l - \alpha_i}{u_i v_l} \\ \frac{\alpha_j - \alpha_m}{u_m v_j} & \frac{\alpha_k - \alpha_m}{u_m v_k} & \frac{\alpha_l - \alpha_m}{u_m v_l} \\ \frac{\alpha_j - \alpha_p}{u_p v_j} & \frac{\alpha_k - \alpha_m}{u_m v_k} & \frac{\alpha_l - \alpha_p}{u_p v_l} \end{bmatrix},$$

where $r+1 \leq j < k < l \leq n$, $1 \leq i < m < p \leq r$. The determinant of this matrix is equal to 0 as may be shown by straightforward calculations.

Lemma 4. *Let* $\mathbf{L} = [L_{ij}]$, $i = 1, 2, \ldots, r$; $j = r+1, r+2, \ldots, n$ *be any* $r \times (n-r)$ *matrix of rank* 2 *such that all* L_{ij} *are nonzero and all square submatrices are non singular. Then there exist all different* β_k, $k = 1, 2, \ldots, n$ *and nonzero* $\widetilde{u}_i$, $i = 1, 2, \ldots, r$ *and* $\widetilde{v}_j$, $j = r+1, r+2, \ldots, n$, *such that*

$$L_{ij} = \frac{\beta_j - \beta_i}{\widetilde{u}_i \widetilde{v}_j}. \tag{9}$$

In other words, a matrix $\mathbf{R} = [R_{ij}] = \left[L_{ij}^{-1}\right]$ *can always be represented as a Generalized Cauchy matrix in the form of Eq. (8).*

Proof. Consider first two rows of the matrix $\mathbf{L}$, i.e. $2 \times (n-r)$ matrix

$$\begin{bmatrix} L_{1,r+!} & L_{1,r+2} & \ldots & L_{1j} & \ldots & L_{1n} \\ L_{2,r+1} & L_{2,r+2} & \ldots & L_{2j} & \ldots & L_{2n} \end{bmatrix}. \tag{10}$$

Choose β_1 and β_2 arbitrarily but so as to be different. Choose nonzero $\widetilde{u}_1$ and $\widetilde{u}_2$ such that for any $j = r+1, r+2, \ldots, n$,

$$\widetilde{u}_1 L_{1j} \neq \widetilde{u}_2 L_{2j}.$$

This is always possible since $(n-r) < q$. Represent L_{1j} and L_{2j} as

$$\begin{aligned} L_{1j} &= \frac{\widetilde{\beta_j}-\widetilde{\beta_1}}{\widetilde{u_1}\widetilde{v_j}}, \\ L_{2j} &= \frac{\widetilde{\beta_j}-\widetilde{\beta_2}}{\widetilde{u_2}\widetilde{v_j}}, \end{aligned} \tag{11}$$

with unknown variables β_j and $\widetilde{v}_j$. The linear system (Eq. (11)) has a unique solution

$$\begin{aligned} \beta_j &= \frac{\begin{vmatrix} \beta_1 & -\widetilde{u}_1 L_{1j} \\ \beta_2 & -\widetilde{u}_2 L_{2j} \end{vmatrix}}{\begin{vmatrix} 1 & -\widetilde{u}_1 L_{1j} \\ 1 & -\widetilde{u}_2 L_{2j} \end{vmatrix}}, \\ \widetilde{v}_j &= \frac{\begin{vmatrix} 1 & \beta_1 \\ 1 & \beta_2 \end{vmatrix}}{\begin{vmatrix} 1 & -\widetilde{u}_1 L_{1j} \\ 1 & -\widetilde{u}_2 L_{2j} \end{vmatrix}}. \end{aligned} \tag{12}$$

Moreover, all β_j are different. If $\beta_j = \beta_k$for some $j \neq k$ then we obtain from Eq. (11) that $\begin{vmatrix} L_{1j} & L_{1k} \\ L_{2j} & L_{2k} \end{vmatrix} = 0$ in contradiction of the Lemma condition. It is also evident that all $\widetilde{v}_j$ are nonzero.

We have proved Lemma for $2 \times (n-r)$ matrices $\mathbf{L}$. In general, add to the matrix of Eq. (10) any third row:

$$\begin{bmatrix} L_{1,r+!} & L_{1,r+2} & \ldots & L_{1j} & \ldots & L_{1n} \\ L_{2,r+1} & L_{2,r+2} & \ldots & L_{2j} & \ldots & L_{2n} \\ L_{s,r+1} & L_{s,r+2} & \ldots & L_{sj} & \ldots & L_{sn} \end{bmatrix}.$$

By the Lemma condition this row is a linear combination of the first two rows, say,

$$\begin{aligned} L_{sj} = a_s L_{1j} + b_s L_{2j} = a_s \frac{\widetilde{\beta_j}-\widetilde{\beta_1}}{\widetilde{u_1}\widetilde{v_j}} + b_s \frac{\widetilde{\beta_j}-\widetilde{\beta_2}}{\widetilde{u_2}\widetilde{v_j}}, \\ s = 3, \ldots r;\ j = r+1, r+2, \ldots, n. \end{aligned} \tag{13}$$

Rewriting L_{sj} as

$$L_{sj} = \frac{\beta_j - \beta_s}{\widetilde{u}_s \widetilde{v} j}$$

with unknown β_s and $\widetilde{u}_s$ we obtain from Eq. (13) that

$$\begin{aligned} \beta_s &= a_s \frac{\beta_1}{\widetilde{u_1}} + b_s \frac{\beta_2}{\widetilde{u_2}}, \\ \frac{1}{\widetilde{u_s}} &= a_s \frac{1}{\widetilde{u_1}} + b_s \frac{1}{\widetilde{u_2}}. \end{aligned} \tag{14}$$

Again, it is easy to show that all β_s are different and all $\widetilde{u}_s$ are nonzero.

Remark. It follows from Eq.'s (12) and (14) that the number of calculations to obtain all $\beta, \widetilde{u}, \widetilde{v}$ from known $\widetilde{u}_1, \widetilde{u}_2, \beta_1, \beta_2, L_{ij}$ is equal to $O(n)$.

Theorem 5. *Let*

$$\mathbf{H}_1 = \begin{bmatrix} z_1 & z_2 & \cdots z_n \\ z_1\alpha_1 & z_2\alpha_2 & \cdots z_n\alpha_n \\ z_1\alpha_1^2 & z_2\alpha_2^2 & \cdots z_n\alpha_n^2 \\ \cdots & \cdots & \cdots\cdots \\ z_1\alpha_1^{r-1} & z_2\alpha_2^{r-1} & \cdots z_n\alpha_n^{r-1} \end{bmatrix}$$

and

$$\mathbf{H}_2 = \begin{bmatrix} w_1 & w_2 & \cdots w_n \\ w_1\beta_1 & w_2\beta_2 & \cdots w_n\beta_n \\ w_1\beta_1^2 & w_2\beta_2^2 & \cdots w_n\beta_n^2 \\ \cdots & \cdots & \cdots\cdots \\ w_1\beta_1^{r-1} & w_2\beta_2^{r-1} & \cdots w_n\beta_n^{r-1} \end{bmatrix}$$

***be two different parity check matrices of the standard form defining the same** GRS code. Then*

$$\frac{z_j}{z_i} \prod_{k=1,k\neq i}^{r} \frac{\alpha_j - \alpha_k}{\alpha_i - \alpha_k} = \frac{w_j}{w_i} \prod_{k=1,k\neq i}^{r} \frac{\beta_j - \beta_k}{\beta_i - \beta_k}$$

or, equivalently

$$\frac{u_i v_j}{\alpha_j - \alpha_i} = \frac{\widetilde{u}_i \widetilde{v}_j}{\beta_j - \beta_i}.$$

Proof. It follows immediately from Lemmas 1 and 4 and Corollaries 2 and 3.

3 The Niederreiter PKC

This Public-Key Cryptosystem is based on Generalized Reed-Solomon codes. The cryptographer A chooses as *Secret Keys* the following:

1. Some $r \times n$ parity check matrix $\mathbf{H}$ of the form in Eq. (1). This matrix provides the Fast Decoding Algorithm for the cryptographer.
2. Some randomly chosen but nonsingular square scrambling matrix $\mathbf{S}$ of order r. This matrix hides any visible structure in the matrix $\mathbf{H}$.

The user calculates and publishes as a Public Key in the public directory the matrix

$$\mathbf{H}_{cr} = \mathbf{SH}.$$

The sending party B fetches $\mathbf{H}_{cr}$ from the public directory and encrypts the plaintext (which is an n-bit message $\overline{\mathbf{m}}$ with Hamming weight $t \leq r/2$) as a syndrome

$$\overline{\mathbf{c}} = \overline{\mathbf{m}}\mathbf{H}_{cr}^t = \overline{\mathbf{m}}\mathbf{H}^t\mathbf{S}^t.$$

Upon receipt of $\overline{\mathbf{c}}$, the legitimate receiver A calculates $\overline{\mathbf{m}}\mathbf{H}^t$ as $\overline{\mathbf{c}}\,(\mathbf{S}^t)^{-1}$ and applies to this vector the Fast Decoding Algorithm, recovering the plaintext $\overline{\mathbf{m}}$.

4 The Sidel'nikov-Shestakov Attack

There exist two kinds of attacks on any cryptosystem.

The opponent E may try to obtain plaintext $\overline{\mathbf{m}}$ immediately from an intercepted ciphertext $\overline{\mathbf{c}}$. This is a *first* approach. It seems that for the Niederreiter PKC this approach is infeasible. This attack is based, in essence, on the exhaustive search of all possible plaintext. Let W denote the Work function, i.e., the number of calculations to break a cryptosystem. Then

$$W_1 \sim nC_n^t.$$

Thus for moderate values of n and t ($n = q \geq 128$, $t \geq 32$) the Work function is rather large to break the cryptosystem.

The opponent E may try instead to obtain *Secret Keys* from the known *Public Key.* This is a *second* approach.

The main idea of the Sidel'nikov-Shestakov attack is finding from the Public Key $\mathbf{H}_{cr} = \mathbf{SH}$ (where $\mathbf{H}$ and $\mathbf{S}$ are unknown to the opponent E) any trapdoors $\mathbf{H}_{tr}$ and $\mathbf{S}_{tr}$ such that

$$\mathbf{H}_{cr} = \mathbf{SH} = \mathbf{S}_{tr}\mathbf{H}_{tr},$$

where

$$\mathbf{H}_{tr} = \begin{bmatrix} w_1 & w_2 & \cdots w_n \\ w_1\beta_1 & w_2\beta_2 & \cdots w_n\beta_n \\ w_1\beta_1^2 & w_2\beta_2^2 & \cdots w_n\beta_n^2 \\ \cdots & \cdots & \cdots\cdots \\ w_1\beta_1^{r-1} & w_2\beta_2^{r-1} & \cdots w_n\beta_n^{r-1} \end{bmatrix}$$

is a matrix of the standard form but w and β may differ from z and α in the secret keys. If the opponent obtains trapdoors he knows the other Fast Decoding Algorithm and can decrypt $\overline{c}$ in just the same manner as the legitimate receiver A.

To get trapdoors the opponent E reduces the matrix $\mathbf{H}_{cr}$ (i.e., the Public Key) to the systematic form and applies constructions of Lemma 4 and Theorem 5 to obtain some w and β. This breaks the *Niederreiter* PKC completely because the Work function is about

$$W_2 = O\left(n^3\right).$$

5 Modification of the Niederreiter PKC

The original Niederreiter PKC can be effectively modified by destroying the very simple structure of the systematic parity check matrix (see Lemma 1 and Corollaries). To realize this idea we change the original Public Key $\mathbf{H}_{cr}$ to $\mathbf{H}_{cr,mod} = \mathbf{H}_{cr} + \mathbf{SX}$, where $\mathbf{X}$ is a matrix of rank 1 and of the same size as $\mathbf{H}_{cr}$. The hiding matrix $\mathbf{X}$ may be chosen as follows:

- Choose any vector $\overline{\mathbf{a}} = (a_1, a_2, \ldots, a_n)$ from any coset of weight $d-1 = 2t = r$ of the given GRS code.

- Calculate a syndrome $\overline{\mathbf{x}} = \overline{\mathbf{a}}\mathbf{H}^t_{cr} = (x_1.x_2, \ldots, x_{d-1})$.
- Calculate $\mathbf{X}$ as

$$\mathbf{X} = \overline{\mathbf{x}}^t \overline{\mathbf{b}},$$

where $\overline{\mathbf{b}} = (b_1, b_2, \ldots, b_n)$ and not all zero elements $b_i \in GF(q)$ are chosen randomly.
- Calculate the modified Public Key as

$$\mathbf{H}_{cr,mod} = \mathbf{S}(\mathbf{H} + \mathbf{X}).$$

- Use as plaintexts all vectors $\overline{\mathbf{m}} = (m_1, m_2, \ldots, m_n)$ of weight less than or equal to $t-1$ (not t as before).

Then a modified ciphertext $\overline{\mathbf{c}}$ is equal to

$$\overline{\mathbf{c}} = \overline{\mathbf{m}}\mathbf{H}^t_{cr,mod} = \overline{\mathbf{m}}\mathbf{H}^t\mathbf{S}^t + \overline{\mathbf{m}}\mathbf{X}^t\mathbf{S}^t = \overline{\mathbf{m}}\mathbf{H}^t\mathbf{S}^t + \lambda(\overline{\mathbf{m}})\overline{\mathbf{x}}\mathbf{S}^t$$

where $\lambda(\overline{\mathbf{m}})$ is an element from the field $GF(q)$.

The legitimate receiver A knows $\overline{\mathbf{x}}$ but not $\lambda(\overline{\mathbf{m}})$ and should decode at most n times to get plaintext $\overline{\mathbf{m}}$. Thus, the decoding complexity is increased not more than n times. This is a disadvantage in the proposed modification.

The opponent E, on the contrary, has to eliminate $\mathbf{X}$ before he can use the Sidel'nikov-Shestakov algorithm on $\mathbf{H}_{cr}$. Since there are many different elements in the coset of the given GRS code with leaders of weight $d-1$ (the entropy of $\mathbf{X}$ is large), it is thought to be a difficult and time-consuming task.

References

1. H. Niederreiter, "Knapsack-Type Cryptosystem and Algebraic Coding Theory," *Probl. Control and Inform. Theory*, vol.15, pp.19-34, 1986.
2. V.M. Sidelnikov and S.O. Shestakov, "On the Insecurity of Cryptosystems Based on Generalized Reed-Solomon Codes," *Discrete Math.*, vol. 1, no. 4, pp. 439-444, 1992.
3. V.M. Sidelnikov and S.O. Shestakov, "On the Cryptosystem Based on Generalized Reed-Solomon Codes," in the Report *Prospective Telecommunication and Integrated Communication Systems*, Institute for Problems of Information Transmission, Russian Academy of Science, Moscow, 1992, pp. 48 - 61 (in Russian).
4. F.J. MacWilliams and N.J.A. Sloane, The Theory of Error-Correcting Codes, North-Holland Publishing Company, New-York, 1977.

Linear Algebra Approach to Secret Sharing Schemes

G.R.Blakley[1] and G.A. Kabatianskii[2]

[1] Texas A&M University, College Station,
TX 77843-3368, USA;
blakley@math.tamu.edu

[2] Institute of Problems for Information Transmission,
Moscow, 101447 Russia;
kaba@ippi.msk.su

Abstract. The problem of secret sharing schemes (SSS) in the case where all sharing functions are linear maps over a finite field is investigated. We evaluate the performance of linear secret sharing schemes using the tools of linear algebra and coding theory. In particular, the nonexistence of an ideal threshold linear SSS for the case where the number of participants is twice as large as the number of possible values of a secret is shown.

1 Introduction

Let us briefly review the problem. n participants and a dealer want to share a secret. The secret is a binary (or q-ary in the general case) sequence of length m. The dealer facilitates the sharing by giving each of the participants one or more sequences, called "shadows" or "shares" (shadows can have different lengths for different participants). The dealer gives shadows in such a way that certain subsets of participants, called allowed (qualified, authorized) coalitions, can recover the secret while all other possible coalitions (nonallowed) can derive no additional (*a posteriori*) information about the value of the secret from the values of their shadows.

This problem was introduced and solved for the case of (n, s)-*threshold* schemes, where allowed coalitions consist at least s participants, in [4, 5]. This result was reproved later through the use of coding theory [6]. Recently the general relationship between linear codes and secret sharing schemes has been found [7, 8, 9]. Following the same techniques, we investigate the relation between sizes of alphabets for secrets and shadows and the number of participants. In particular, we show that the ideal linear perfect secret sharing scheme exists if and only if the corresponding MDS code [1] exists, which leads to the nonexistence of schemes for the case where $n > 2q - 2$.

In fact, the coding theoretic approach can be reformulated as the vector space construction, which was introduced in [2]. We propose a more general approach in which general linear transformations can be used; we are not restricted to linear functionals. The problem of linear SSS can then be stated in the following form. For given integers n and m, find (or prove the nonexistence of) a family of subspaces of an n-dimensional vector space over a finite field with two properties:

- The linear span of allowed collections of subspaces contains a fixed m-dimensional subspace;
- The linear span of any collection of subspaces corresponding to a nonallowed coalition of participant has only the zero vector in common with this m-dimensional subspace.

2 Secret Sharing Schemes in General

Let $\{1, 2, \ldots, n\}$ be a set of participants. As usual, $\mathcal{P}(Y)$ is the power set of the set Y. A set Γ of subsets of $\{1, 2, \ldots, n\}$ (formally $\Gamma \subseteq \mathcal{P}(\{1, 2, \ldots, n\})$ is said to be an *access structure* if the monotone property holds, *i.e.* if $A \subset B, A \in \Gamma$ then $B \in \Gamma$. Denote by $\overline{\Gamma}$ the complementary set to the set Γ, *i.e.* $\overline{\Gamma} = \{A \subseteq \{1, 2, \ldots, n\} : A \notin \Gamma\}$.

Let K be a set of possible secrets and p is a probability distribution on it. The dealer distributes a secret $k \in K$, which is selected with probability $p(k)$, by giving the i-th participant a shadow $x(i)$ from a set $X(i)$. In this implementation he uses an auxiliary finite set E with a probability distribution function (pdf) P and a mapping $\mathcal{F}$, where

$$\mathcal{F} : E \rightarrow K \times X(1) \times \cdots \times X(n) \tag{1}$$

Suppose that E, P and $\mathcal{F}$ generate the *SSS*. This means that the dealer chooses an element $e \in E$ at random (*i.e.* in a manner obeying the pdf P) and, hence, produces a secret k and shadows $x(i)$ in accordance with the mapping $\mathcal{F}$.

$$\mathcal{F}(e) = (f(e), f_1(e), \ldots, f_n(e)) = (k, x(1), x(2), \ldots, x(n)) \tag{2}$$

Thus a choice of $e \in E$ entails a unique choice of $k, x(1), x(2), \ldots, x(n)$. Of course, we demand that the mapping $\mathcal{F}$ and the pdf P induce on the set K the desired pdf p.

Definition 1. A triple $(E, P, \mathcal{F})$ is said to generate a perfect *SSS* for the access structure Γ if the following conditions hold:

1. For every $A \in \Gamma$, every $b \in K$, and every $\nu(a) \in X(a)$ where $a \in A$,

$$\mathbf{Pr}(k = b \mid x(a) = \nu(a) \text{ for } a \in A) \in \{0, 1\}$$

2. For every $A \in \overline{\Gamma}$, every $b \in K$, and every $\nu(a) \in X(a)$ where $a \in A$,

$$\mathbf{Pr}(k = b \mid x(a) = \nu(a) \quad \text{for } a \in A) = \mathbf{Pr}(k = b)$$

In other words, the members of a qualified set A of participants can uniquely determine the secret k by the "list" $\{(a, x(a)) : a \in A\}$ of their shadows, and the members of a nonqualified set A of participants find that their *a posteriori* estimate of k's value is the same as their *a priori* estimate.

Remark. For an arbitrary set Γ of subsets of $\{1, 2, \ldots, n\}$, *i.e.* $\Gamma \subseteq \mathcal{P}(\{1, 2, \ldots, n\})$, an element A is said to be minimal (maximal) if there is no element $B \in \Gamma$ such that $B \subset A$ ($B \supset A$). Let $\partial\Gamma$ denote a set of minimal elements of Γ and $\nabla\Gamma$ denote a set of maximal elements of Γ. It is obvious that replacing Γ and $\overline{\Gamma}$ with $\partial\Gamma$ and $\nabla\overline{\Gamma}$ within Definition 1 gives an equivalent definition.

The *rate* R of SSS is defined (see [11]) as the ratio

$$\frac{\log|K|}{\max\{\log|X(i)|\}}$$

It is known that for perfect SSS, $|X(i) \geq |K||$ for all i, and hence $R \leq 1$. A perfect SSS is said to be *ideal* if $R = 1$, *i.e.* if $|X(i)| = |K|$ for all i.

3 Linear Secret Sharing Schemes: Matrix Formulation

Let all of the above considered sets, *i.e.* $E, K, X(1), X(2), \ldots, X(n)$, be vector spaces over some finite field $GF(q)$, and let $\mathcal{F}$ be a linear transformation. Suppose in addition that p and P are the uniform pdf's which correspond to the counting measure on the finite sets K and E respectively. Such an SSS is said to be *linear*.

Denote by r the dimension of E and by $m_0, m_1, \ldots, m_n$ the dimensions of spaces $X(0), X(1), X(2), \ldots, X(n)$, where we set $X_0 = K$ for notational convenience. Let us start our consideration with the case $m_0 = m_1 = \ldots = m_n = 1$. We will refer to such an SSS as a *one-dimensional* linear SSS. To the linear transformation $\mathcal{F}$ there corresponds an r by $n+1$ matrix H which can be written in partitioned form as a row of $n+1$ columns of height r (*i.e.* r by 1 matrices) in the following way.

$$H = [\ h_0 \mid h_1 \mid h_2 \mid \ \ldots \ \mid h_n\ ] \tag{3}$$

Hence, $F(e) = eH = (x(0), x(1), \ldots, x(n))$, where the element $x(i)$ is of the form $x(i) = eh_i$ the matrix product of the 1 by r matrix (row vector) e sitting on the left of the r by 1 matrix (column vector) h_i. In other words, the element $x(i)$ is the inner product of the vectors e and h_i, which is denoted as usual by $x(i) = (e, h_i)$. We will say that the vectors $h_0, h_1, h_2, \ldots, h_n$ generate a one-dimensional linear SSS. Let $SpanT$ denote, as usual, the linear subspace of the vector space E consisting of all vectors which can be represented as linear combination of vectors of the set T.

Lemma 2 [2]. *A one-dimensional linear SSS generated by vectors $h_0, h_1, \ldots, h_n$ is perfect if and only if the following properties hold:*

1. *For every $A \in \Gamma$ the vector $h_0 \in Span\{h_a : a \in A\}$, i.e. the vector h_0 is a linear combination of vectors h_a, $a \in A$;*

2. *For every $A \in \overline{\Gamma}$ the vector $h_0 \notin Span\{h_a : a \in A\}$.*

Proof. Let A be an arbitrary subset of the set $\{1,2,\ldots,n\}$. Consider the case $h_0 \in \mathrm{Span}\{h_a : a \in A\}$. Then $h_0 = \sum_{a \in A} \lambda_a h_a$ for some coefficients $\lambda_a \in .GF(q)$ and therefore $k = (e, h_0) = \sum_{a \in A} \lambda_a (e, h_a) = \sum_{a \in A} \lambda_a x(a)$. Hence members of the set A can recover the secret k as it is a linear combination of the shadows $x(a)$ with the same coefficients λ_a.

Consider the case $h_0 \notin \mathrm{Span}\{h_a : a \in A\}$. We will show that the cardinality of the set

$$E(b, \{\nu(a) : a \in A\}) = \{e \in E : (e, h_0) = b,\ (e, h_a) = \nu(a) \forall a \in A\}$$

does not depend on b. Hence the *a posteriori* probability of k is uniform, *i.e.* $\mathbf{Pr}(k = b \mid x(a) = \nu(a)\ \forall a \in A) = 1/q)$, and coincides with its *a priori* probability $\mathbf{Pr}(k = b)$ (see condition 2 of definition 1). To prove this we need only the good old Kronecker-Capelli theorem. Namely, a system of inhomogeneous linear equations is compatible if and only if the rank of the augmented matrix of the system equals the rank of the coefficient matrix. Denote by $H_{0 \cup A}$ and $H_{0 \cup A}^{aug}$ the coefficient matrix and the augmented matrix of the system defining the set $E(b, \{\nu(a) : a \in A\})$:

$$\begin{cases} eh_0 = b \\ eh_a = \nu(a), \quad \forall\, a \in A \end{cases} \tag{4}$$

Let H_A and H_A^{aug} be corresponding matrices of that system but without the first equation $(e, h_0) = b$. Then $rank(H_{0 \cup A}) = 1 + rank(H_A)$ and $rank(H_{0 \cup A}^{aug}) = 1 + rank(H_A^{aug})$ as $h_0 \notin \mathrm{Span}\{h_a : a \in A\}$. Hence the validity of the condition of the Kronecker-Capelli theorem does not depend on b. Together with the remark that the number of solutions of a compatible system is the same for any constant terms of the system (and equals the number of solutions of the corresponding homogeneous system), it gives the proof of the "if" part. The necessity of the conditions of the Lemma follows immediately from the above consideration, because there are only two possibilities for the vector h_0 - it either belongs or does not belong to the subspace $\mathrm{Span}\{h_a : a \in A\}$.□

Consider the case of general linear SSS. To the linear transformation $\mathcal{F}$ there corresponds now an r by N matrix H , where $N = m_0 + m_1 + \cdots + m_n$. The matrix H can be written in the following way:

$$H = [\ H_0 \mid H_1 \mid \ldots \mid H_n\ ] \tag{5}$$

where H_i is an r by m_i matrix and can be represented as a set (or as a row) of m_i column vectors $h_{i,j}$, $j = \overline{1, m_i}$, *i.e.*

$$H_i = [\ h_{i,1} \mid h_{i,2} \mid \ldots \mid h_{i,m_i}\ ] \tag{6}$$

Let $\mathcal{H}_i$ be the vector space generated by vectors $h_{i,j}$, $j = \overline{1, m_i}$, *i.e.* $\mathcal{H}_i = \text{Span}\{h_{i,j} : j = \overline{1, m_i}\}$. Now we can formulate an analogue of Lemma 2 for the general case.

Theorem 3. *A linear SSS generated by matrices $H_0, H_1, \ldots, H_n$ is perfect if and only if the following properties hold:*

1. *For every $A \in \Gamma$ the vector space $\mathcal{H}_0$ is a subspace of Span$\{\mathcal{H}_a : a \in A\}$*
2. *For every $A \in \overline{\Gamma}$ the vector space $\mathcal{H}_0$ has only the zero vector in common with the vector space Span$\{\mathcal{H}_a : a \in A\}$.*

Sketch of Proof. There are three possible cases:

- $\mathcal{H}_0 \subseteq \text{Span}\{\mathcal{H}_a : a \in A\}$

- $\mathcal{H}_0 \cap \text{Span}\{\mathcal{H}_a : a \in A\} = \mathbf{0}$

- $0 < dim(\mathcal{H}_0 \cap \text{Span}\{\mathcal{H}_a : a \in A\}) < dim\mathcal{H}_0$

The analyses of the first two cases , which are cases 1. and 2. from the statement, are the same as in Lemma 2. Namely, it follows from 1. that all vectors $h_{o,j}$ can be represented as linear combinations of the vectors $h_{a,j}$, $a \in A$. Hence the members of the qualified set A can recover the secret k, since its components are linear combinations of the shadows $x(a)$ with the same coefficients .

Consider a set

$$E(\mathbf{b}, \{\nu(a) : a \in A\}) = \{e \in E : eH_0 = \mathbf{b},\ eH_a = \nu(a)\ \forall a \in A\}$$

and a system of linear equations defining this set :

$$\begin{cases} eH_0 = \mathbf{b} \\ eH_a = \nu(a), \quad \forall\, a \in A \end{cases} \tag{7}$$

It follows from condition 2. that the ranks of the augmented and coefficient matrices coincide for this system. Therefore the number of solutions of the system does not depend on $\mathbf{b}$, and property 2 of the Definition is satisfied in this case.

For the third case we should consider again the set $E(\mathbf{b}, \{\nu(a) : a \in A\})$ and the corresponding system. Choose a basis $\tilde{h}_{0,1}, \tilde{h}_{i,2}, \ldots, \tilde{h}_{0,m_0}$ of the vector space $\mathcal{H}_0$ in such a way that the first l vectors are a basis of the vector space $\mathcal{H}_0 \cap \text{Span}\{\mathcal{H}_a : a \in A\}$. Replace the first group of equations of the system, namely, $eH_0 = \mathbf{b}$, with an equivalent group of equations $e\tilde{H}_0 = \tilde{\mathbf{b}}$. The number of solutions of the new system does not depend on the last $m_0 - l$ coordinates of the constant term $\tilde{\mathbf{b}}$. On the other hand, row vectors $\tilde{h}_{0,1}, \tilde{h}_{i,2}, \ldots, \tilde{h}_{0,l}$ of the coefficients of the new system are linear combinations of the vectors $h_{a,j}$, $a \in A$. Therefore the system is compatible iff the first l coordinates of the constant term $\tilde{\mathbf{b}}$ can be represented as the same linear combinations of the corresponding coordinates of vectors $\nu(a) : a \in A$. Hence, members of the set A can recover exactly l (from m) q-ary digits of the secret. □

Remark. According to the remark after Definition 1, the Theorem and Lemma are true if we replace Γ and $\overline{\Gamma}$ with $\partial\Gamma$ and $\nabla\overline{\Gamma}$.

Example 1. Consider $n = 4$ and an access structure Γ with $\partial\Gamma = \{\{1,2\},\{2,3\},\{3,4\}\}$. It is the first example of an access structure which has no *ideal* realization, and even more, for every perfect realization its rate $R \leq 2/3$ (see [10]). The following choice of $E = GF(q)^6$ and matrices $H_0, H_1, \ldots, H_4$ gives a perfect SSS with rate $R = 2/3$ (see Example 7.1 in[11]):

$$\mathbf{H}_0 = \begin{Vmatrix} 1\,0 \\ 0\,1 \\ 0\,0 \\ 0\,0 \\ 0\,0 \\ 0\,0 \end{Vmatrix}, \mathbf{H}_1 = \begin{Vmatrix} 0\,0 \\ 0\,0 \\ 1\,0 \\ 0\,1 \\ 0\,0 \\ 0\,0 \end{Vmatrix}, \mathbf{H}_2 = \begin{Vmatrix} 1\,0\,0 \\ 0\,1\,0 \\ 1\,0\,0 \\ 0\,1\,0 \\ 0\,0\,1 \\ 0\,0\,0 \end{Vmatrix}, \mathbf{H}_3 = \begin{Vmatrix} 0\,0\,1 \\ 0\,0\,0 \\ 0\,0\,0 \\ 0\,1\,0 \\ 0\,0\,1 \\ 1\,0\,0 \end{Vmatrix}, \mathbf{H}_4 = \begin{Vmatrix} 0\,0 \\ 0\,1 \\ 0\,0 \\ 0\,0 \\ 1\,0 \\ 0\,1 \end{Vmatrix}.$$

4 One-Dimensional Linear SSS and Linear Codes

Consider an r by $n+1$ matrix H, which generates a perfect one-dimensional linear SSS for an access structure Γ, as a parity check matrix of a linear q-ary code V of length $n+1$. We enumerate positions of codewords as $0, 1, \ldots, n$. It is simple to show that property 1., *i.e.* $h_0 \in \text{Span}\{h_a : a \in A\}$, is equivalent to the existence of a codeword v with $v_o = 1$ and $Supp(v) \subseteq A$, where $Supp(v) = \{i : v_i \neq 0\}$ is a support of the vector v. Following [7], define a codeword v to be ***minimal*** if there is no codeword u with $Supp(u) \subset Supp(v)$, *i.e.* v "covers" no other codewords. It is clear that if two minimal codewords have the same support then they differ only by a multiplicative constant. Hence a set $\partial\Gamma$ coincides with a set of supports of minimal codewords of the code V, in which the zero coordinate equals 1.

One of the most important questions that arises is "What kind of access structures can be realized as one-dimensional linear SSS ?" This question can be reformulated now in the following form - "What kind of collections of subsets of the coordinate set $0, 1, \ldots, n$ are realized as supports of minimal codewords of a linear code ?"

Up to now there are more questions than results regarding the way coding theory can be applied to the investigation of SSS . The following simple result shows that there are no one-dimensional linear threshold SSS except those generated by parity check matrices of MDS codes. Recall (see [1]) that a linear code V is said to be MDS code if $d(V) = 1 + rankH$, where $d(V)$ is the minimal Hamming code distance of the code V and H is its parity check matrix. This is equivalent to the property that any $rankH$ columns of the matrix H are linear independent.

Theorem 4. *A matrix H over a finite field $GF(q)$ generates an ideal perfect (n, s)-threshold SSS if and only if H is a parity check matrix for a q-ary linear MDS code of length $n+1$ with minimal code distance $d = s+1$.*

Proof. The "if" part of the statement was proved in [6] and follows immediately from lemma 2.

Consider the "only if" part. Let the matrix H generate a perfect (n, s)-threshold SSS. There are s linear dependent columns $h_{i_1}, \ldots, h_{i_s}$ of the matrix H. On the other hand, the vector h_o can be represented as a linear combination of the vectors $h_{i_1}, \ldots, h_{i_s}$ (condition 1. of lemma ??). It follows that the vector h_o can be represented as a linear combination of some $s-1$ vectors from $h_{i_1}, \ldots, h_{i_s}$. But this contradicts condition 2' of lemma 2. We must now prove that $rank H = s$. Consider any $s+1$ vectors $h_{i_1}, \ldots, h_{i_{s+1}}$. The vector h_0 can be represented as a linear combination of the vectors $h_{i_1}, \ldots, h_{i_s}$ and as a linear combination of the vectors $h_{i_2}, \ldots, h_{i_{s+1}}$ (condition 1. of lemma 2). Hence their difference gives a linear dependence between all vectors $h_{i_1}, \ldots, h_{i_{s+1}}$. □

The (n, s)-threshold SSS is said to be nontrivial if $1 < s < n$. It is obvious that for given q, trivial perfect threshold SSS exist for any n. The following inequalities are known for the length N of nontrivial (*i.e.* $2 < d < N$) MDS (N, k)codes: $N < q + k$ and $N < q + r$, where $r = N - k$. Hence, $N < 2q - 1$ and there is a popular conjecture [1] that $N \leq q + 1$ except for the case $q = 2^i$, $r = 3$, $N = q + 2$. Together with Theorem 4 it gives

Corollary 5. *There is no one-dimensional linear perfect realization of a non-trivial threshold SSS if $n > 2q - 3$, where n is the number of participants, and q is the number of possible values of the secret.*

References

1. MacWilliams, F.J., and Sloane, N.J.A., *The theory of error-correcting codes*, North-Holland Publishing Company, 1977.
2. Brickell E.F., "Some ideal secret sharing schemes", *J. Combin. Math. and Combin. Comput.*, vol. 9, pp. 105-113, 1989.
3. Simmons G.J., "An introduction to shared secret and/or shared control schemes and their applications". In *Contemporary Cryptology:the Science of Information Integrity*, IEEE Press, N.Y., pp.441-497, 1992.
4. Blakley G.R., "Safeguarding Cryptographic Keys". In *Proc. AFIPS 1979 Natl.Computer Conference*, NY, vol. 48, pp. 313-317, 1979.
5. Shamir A., "How to share a secret", *Communications of ACM*, vol. 22, pp. 612-613, 1979.
6. McEliece R.J., Sarwate D.V.," On secret sharing and Reed-Solomon codes", *Communications of ACM* , vol. 24, pp. 583-584, 1981.
7. Massey J.L., "Minimal codewords and secret sharing". In *Proc. Sixth Joint Swedish-Russian Workshop on Information theory*, Molle, Sweden, pp. 276-279, 1993.
8. Bertillson M., "Linear codes and secret sharing". Linkoping Studies in Science and Technology. Dissertation, No.299, 1993.
9. Blakley G.R., Kabatianskii G.A., "Linear algebra approach to secret sharing schemes". In *PreProceedings of Workshop on Information Protection*, Moscow, 1993.

10. Capocelli R.M., De Santis A., Gargano L. and Vacaro U."On the size of shares for secret sharing schemes". In *Lecture Notes in Computer Science*, Springer-Verlag, v. 576, pp. 101-113, 1992.
11. Stinson D.R., "An explication of secret sharing schemes ",*Designs,Codes and Cryptography*, vol. 2, pp. 357-390, 1992.

Generalizations of the Griesmer Bound *

Tor Helleseth, Torleiv Kløve, Øyvind Ytrehus

Department of Informatics, University of Bergen, HIB, N-5020 Bergen, Norway

Abstract. Various generalizations of the Griesmer bound to minimum support weights are given. The chain condition for codes of lengths at most 2 above the Griesmer bound is discussed.

1 Introduction

For any code D, $\chi(D)$, the *support of* D, is the set of positions where not all the codewords of D are zero, and $w_S(D)$, the *support weight* of D, is the size of $\chi(D)$. For an $[n,k]$ code C and any r, where $1 \le r \le k$, *the r-th minimum support weight* is defined by

$$d_r = d_r(C) = \min\{w_S(D) \mid D \text{ is an } (n,r) \text{ subcode of } C\} \ .$$

In particular, the minimum distance of C is d_1. The *weight hierarchy* of C is the set $\{d_1, d_2, \cdots, d_k\}$. The r-th minimum support weight is also known as *r-th generalized Hamming weight* [30]. The weight hierarchy has been called the *length/distance profile* [8].

The weight hierarchy has been studied by a number of researchers in the last couple of years. In the bibliography we have listed all papers, known to us, that deal with support weights and related questions.

There are a number of results which can be seen as generalizations of the Griesmer bound. In this paper we state these and discuss proofs.

Further, we use the following notation:

$$\mathcal{D}_r = \{D \mid D \text{ is a subspace of } C \text{ of dimension } r \text{ such that } w_S(D) = d_r(C)\} \ .$$

In the terminology of Wei and Yang [31], we say that C satisfies the *chain condition* if there exist $D_r \in \mathcal{D}_r$ for $1 \le r \le k$ such that

$$D_1 \subseteq D_2 \subseteq \cdots \subseteq D_k,$$

and we call $\{D_1, D_2, \cdots, D_k\}$ a *chain* for C. We discuss the chain condition for codes of lengths at most 2 above the Griesmer bound.

* Research supported by the Norwegian Research Council

2 The Excess Sequences of a Code

Consider an $[n, k]$ code C. For $r \geq s$ let

$$g_s(r,d) = d + \sum_{i=1}^{r-s} \left\lceil \frac{d}{2^i(2^s-1)} \right\rceil .$$

Define the (r, s)-*excess* of C by

$$e_s(r) = e_s(C,r) = d_r(C) - g_s(r,d_s) .$$

Note that $e_s(s) = 0$. The s-th *excess sequence* of C is the sequence $(e_s(s), e_s(s+1), \ldots, e_s(k))$.

Theorem 1. *Let C be an $[n, k]$ code with weight hierarchy $\{d_1, d_2, \cdots, d_k\}$. For $1 \leq s \leq k$ we have*

$$0 = e_s(s) \leq e_s(s+1) \leq e_s(s+2) \leq \cdots \leq e_s(k) .$$

This theorem has a number of corollaries.

Corollary 2. *For $s \leq r \leq k$ we have $e_s(r) \geq 0$, that is,*

$$d_r \geq g_s(r, d_s) .$$

Since $n \geq d_k$ we get in particular the following bound.

Corollary 3. *For $1 \leq s \leq k$ we have*

$$n \geq g_s(k, d_s) .$$

Another special case is obtained putting $s = 1$ in Corollary 2.

Corollary 4. *For $1 \leq r \leq k$ we have*

$$d_r \geq g_1(r, d_1) .$$

This result is sometimes referred to as the Wei-Griesmer bound. Finally, if we put $s = 1$ in Corollary 3 we get the Griesmer bound:

Corollary 5. *We have*

$$n \geq g_1(k, d_1) .$$

Theorem 1 in its full generality is stated here for the first time. The special case with $s = 1$ was given in [15] and [16]. Corollaries 2 and 3 were given in [13]. Corollary 4 was used implicitly by Wei [30] and Corollary 5 is a classical result due to Griesmer [11].

In [13] we gave the following two lemmata. Both are simple to prove, and we state them here without proof.

Lemma 6. *If $1 \leq r < k$, then*

$$d_{r+1} \geq d_r + \frac{d_r}{2^{r+1}-2}.$$

Lemma 7. *For $r \geq s$ we have*

$$g_s(r+1,d) = g_s(r,d) + \left\lceil \frac{g_s(r,d)}{2^{r+1}-2} \right\rceil .$$

The proof of Theorem 1 is by induction. First

$$e_s(s) = d_s - g_s(s,d_s) = d_s - d_s = 0$$

by the definition of g_s. Next, by Lemmata 6 and 7, and the fact that $d_{r+1} - d_r$ is an integer, we get

$$\begin{aligned} e_s(r+1) - e_s(r) &= d_{r+1} - g_s(r+1,d_s) - d_r + g_s(r,d_s) \\ &\geq \left\lceil \frac{d_r}{2^{r+1}-2} \right\rceil - \left\lceil \frac{g_s(r,d_s)}{2^{r+1}-2} \right\rceil \geq 0. \end{aligned}$$

□

We note that, in particular, we get a new proof of the Griesmer bound. Usually, the Griesmer bound is proved using residue codes. We can give a similar direct proof of Corollaries 2 and 3, using only Lemma 6 and not Lemma 7. We first prove Corollary 3.

Let C be an $[n,k]$ code, and let D be a subcode of dimension s and support weight d_s. Consider the residue code E, that is the code obtained from C by puncturing the positions in $\chi(D)$. This is a code of length $n' = n - d_s$ and dimension $k' = k - s$. Its minimum distance d' is

$$\begin{aligned} d' &= \min\{w_S(D') - d_s \mid D \subset D' \subseteq C \text{ and } dim(D') = s+1\} \\ &\geq d_{s+1} - d_s \geq \frac{d_s}{2^{s+1}-2} \end{aligned}$$

by Lemma 6. By the Griesmer bound

$$n - d_s \geq \sum_{i=0}^{k-s-1} \left\lceil \frac{d'}{2^i} \right\rceil \geq \sum_{i=0}^{k-s-1} \left\lceil \frac{d_s}{2^i(2^{s+1}-2)} \right\rceil = \sum_{i=1}^{k-s} \left\lceil \frac{d_s}{2^i(2^s-1)} \right\rceil$$

and Corollary 3 follows. □

To prove Corollary 2, first choose a subcode D of C of support weight d_r and dimension r. Next, let C' be the $[d_r, r]$ code obtained from C by puncturing the positions not in $\chi(D)$. Then $d_s(C') \geq d_s(C)$. By Corollary 3

$$d_r \geq g_s(r, d_s(C')) \geq g_s(r, d_s(C)) .$$

□

Remark. The results above generalize to the non-binary case without problems. For codes over $GF(q)$, the analogue of Lemma 6 is

$$d_{r+1} \geq d_r + \frac{d_r(q-1)}{q^{r+1}-q} ,$$

the generalization of g_s is

$$g_s(r,d) = d + \sum_{i=1}^{r-s} \left\lceil \frac{d(q-1)}{q^i(q^s-1)} \right\rceil ,$$

and Lemma 7 generalizes to

$$g_s(r+1,d) = g_s(r,d) + \left\lceil \frac{g_s(r,d)(q-1)}{q^{r+1}-q} \right\rceil .$$

The proofs of Lemmata 6 and 7 given in [13] as well as all proofs above carry over without any change to the non-binary case.

Theorem 8. *If C is an $[n,k,d]$ code, and $1 \leq s < r \leq k$, then*

$$e_s(k) - e_s(r) \geq \sum_{i=1}^{k-r} \left\{ \left\lceil \frac{\delta_s(r) + e_s(r) - e_s(r-1)}{2^i} \right\rceil - \left\lceil \frac{\delta_s(r)}{2^i} \right\rceil \right\}$$

where $\delta_s(r) = \left\lceil \frac{d_s}{2^{r-s}(2^s-1)} \right\rceil$.

Proof. Let $D \in \mathcal{D}_{r-1}$ and let E be the $[n',k',d']$ code obtained from C by puncturing the positions in $\chi(D)$ and the positions not in $\chi(C)$. Then

$$\begin{aligned} n' &= d_k - w_S(D) \\ &= e_s(k) + d_s + \sum_{i=1}^{k-s} \left\lceil \frac{d_s}{2^i(2^s-1)} \right\rceil - e_s(r-1) - d_s - \sum_{i=1}^{r-1-s} \left\lceil \frac{d_s}{2^i(2^s-1)} \right\rceil \\ &= e_s(k) - e_s(r-1) + \sum_{i=r-s}^{k-s} \left\lceil \frac{d_s}{2^i(2^s-1)} \right\rceil \\ &= e_s(k) - e_s(r-1) + \delta_s(r) + \sum_{i=1}^{k-r} \left\lceil \frac{\delta_s(r)}{2^i} \right\rceil \end{aligned}$$

and $k' = k-(r-1) = k-r+1$. Further, if $\bar{x}$ is a code word in C of weight d' when constrained to E, and $D' = \langle \bar{x}, D \rangle$ is the subcode of C generated by $\bar{x}$ and D, then D' has dimension r and so

$$d_r \leq w_S(D') = d_{r-1} + d' .$$

Hence

$$d' \geq d_r - d_{r-1} = \delta_s(r) + e_s(r) - e_s(r-1) .$$

By the ordinary Griesmer bound we get

$$e_s(k) - e_s(r-1) + \delta_s(r) + \sum_{i=1}^{k-r} \left\lceil \frac{\delta_s(r)}{2^i} \right\rceil$$

$$\geq \delta_s(r) + e_s(r) - e_s(r-1) + \sum_{i=1}^{k-r} \left\lceil \frac{\delta_s(r) + e_s(r) - e_s(r-1)}{2^i} \right\rceil$$

and the theorem follows. □

Next, we will determine the cases when the expression in the right hand of the inequality in Theorem 8 is small (0 or 1). Let

$$\sigma_{srl} = \left\lceil \frac{\delta_s(r) + e_s(r) - e_s(r-1)}{2^l} \right\rceil - \left\lceil \frac{\delta_s(r)}{2^l} \right\rceil ,$$

and

$$\rho_{sr} = \sum_{l=1}^{k-r} \sigma_{srl} .$$

Clearly, if $e_s(r) = e_s(r-1)$, then $\rho_{sr} = 0$.

Lemma 9. *If $e_s(r) > e_s(r-1)$, then $\rho_{sr} = 0$ if and only if*
(a) $r = k$,
or (b) $1 \leq r \leq k-1$, $e_s(r) - e_s(r-1) = 1$, *and $\delta_s(r)$ is odd.*

Proof. Obviously, $\rho_{sk} = 0$. If $r < k$, then it is necessary and sufficient that $\sigma_{sr1} = 0$ and this is equivalent to the condition given in the lemma. □

Lemma 10. *We have $\rho_{sr} = 1$ if and only if*
(a) $r = k-1$, $e_s(r) - e_s(r-1) \in \{2, 3\}$, *and $\delta_s(r)$ is odd,*
or (b) $r = k-1$, $e_s(r) - e_s(r-1) \in \{1, 2\}$, *and $\delta_s(r)$ is even,*
or (c) $2 \leq r \leq k-2$, $e_s(r) - e_s(r-1) \in \{2, 3\}$, *and* $\delta_s(r) \equiv 1 \pmod 4$,
or (d) $2 \leq r \leq k-2$, $e_s(r) - e_s(r-1) \in \{1, 2\}$, *and* $\delta_s(r) \equiv 2 \pmod 4$.

Proof. If $r = k-1$ it is necessary and sufficient that $\sigma_{sr1} = 1$. This gives the first two cases in the lemma. If $r < k-1$ it is necessary and sufficient that $\sigma_{sr1} = 1$ and $\sigma_{sr2} = 0$. This gives the last two cases in the lemma. □

3 Some General Results on Chains

¿From now on we restrict ourselves to the case $s = 1$. For convenience we write $e(r)$ for $e_1(r)$ and $\delta(r)$ for $\delta_1(r)$.

Lemma 11. *Let C be an $[n,k,d]$ code. Let r be maximal such that we can find $D_i \in \mathcal{D}_i$ for $1 \le i < r$ such that*

$$D_1 \subseteq D_2 \subseteq \cdots \subseteq D_{r-1}.$$

If $r \le k$, then

$$e(k) - e(r) \ge 1 + \sum_{i=1}^{k-r} \left\{ \left\lceil \frac{1 + \delta(r) + e(r) - e(r-1)}{2^i} \right\rceil - \left\lceil \frac{\delta(r)}{2^i} \right\rceil \right\} .$$

Proof. By the definition of r, we have $w_S(\langle D_{r-1}, \bar{x} \rangle) > d_r$ for any $\bar{x} \in C \setminus D_{r-1}$. Using the same construction as in the proof of Theorem 8 we get

$$d' \ge 1 + d_r - d_{r-1} = 1 + \delta(r) + e(r) - e(r-1) .$$

The rest of the proof is similar to the proof of Theorem 8. □

Lemma 12. *Let C be an $[n,k,d]$ code. If there exists an l where $2 \le l < k$ such that $e(l+1) = e(l)$ and*
either $e(l-1) < e(l) - 1$
or $e(l-1) = e(l) - 1$ and $\delta(l)$ is even,
then C does not satisfy the chain condition.

Proof. Suppose that C satisfies the chain condition, and let $\{D_1, D_2, \cdots, D_k\}$ be a chain for C. Then $\{D_1, D_2, \cdots, D_{l+1}\}$ is a chain for D_{l+1}. Applying Theorem 8 to D_{l+1} we get

$$0 = e(l+1) - e(l) \ge \left\lceil \frac{\delta(l) + e(l) - e(l-1)}{2} \right\rceil - \left\lceil \frac{\delta(l)}{2} \right\rceil \ge 0$$

which gives a contradiction in both cases. □

For an $[n,k,d]$ code C with generator matrix G, let C^* be the $[n+2^k-1, k, d+2^{k-1}]$ code generated by $(G|S_k)$, where S_k is a $k \times (2^k - 1)$ matrix having as columns all the non-zero vectors in $GF(2)^k$. For any subcode D of C, let D^* be the corresponding subcode of C^*. Then

$$w_S(D^*) = w_S(D) + 2^k - 2^{k-dim(D)} .$$

Hence

$$d_r(C^*) = d_r(C) + 2^k - 2^{k-r}$$

for $1 \le r \le k$. Since $g_1(r, d+2^{k-1}) = g_1(r,d) + 2^k - 2^{k-r}$, we have

$$e(r)(C^*) = e(r)(C)$$

for $1 \le r \le k$. Further, if $D \subseteq E$, then $D^* \subseteq E^*$. Hence, we have proved the following lemma.

Lemma 13. *If there exists an $[n,k,d]$ code C with excess sequence $(e(1), e(2), \cdots, e(k))$, then there exists an $[n+2^k-1, k, d+2^{k-1}]$ code C^* with the same excess sequence. Moreover, if C satisfies the chain condition, then so does C^*.*

4 Codes With $e(k) \leq 2$

First we show that if $e(k) \leq 1$, then C satisfies the chain condition.

Theorem 14. *If C is a binary $[n,k,d]$ code and $e(k) = 0$, then C satisfies the chain condition.*

Proof. By Theorem 1 we have $e(r) = 0$ for $1 \leq r \leq k$. Hence, for all r, $1 < r \leq k$, we have

$$e(k) - e(r) = 0 \leq \sum_{l=1}^{k-r} \left\{ \left\lceil \frac{1+\delta(r)}{2^l} \right\rceil - \left\lceil \frac{\delta(r)}{2^l} \right\rceil \right\}$$

$$= \sum_{l=1}^{k-r} \left\{ \left\lceil \frac{1+\delta(r)+e(r)-e(r-1)}{2^l} \right\rceil - \left\lceil \frac{\delta(r)}{2^l} \right\rceil \right\}.$$

By Lemma 11, C satisfies the chain condition. □

Theorem 15. *If C is a binary $[n,k,d]$ code and $e(k) = 1$, then C satisfies the chain condition.*

Proof. Let r be as in Lemma 11. Since $e(k) = 1$, we get, by Lemma 11,

$$1 - e(r) \geq 1 + \sum_{l=1}^{k-r} \left\{ \left\lceil \frac{1+\delta(r)+e(r)-e(r-1)}{2^l} \right\rceil - \left\lceil \frac{\delta(r)}{2^l} \right\rceil \right\} \geq 1 .$$

Hence $e(r) = 0$. However, by Theorem 14 this implies that there exist subcodes $D_1^*, D_2^*, \cdots, D_r^*$ of D_r (and hence of C) such that $D_i^* \in \mathcal{D}_i$ and

$$D_1^* \subseteq D_2^* \subseteq \cdots \subseteq D_r^*.$$

This contradicts the maximality of r. □

For any code C, define

$$R_j(C) = \min\{r \mid e(r) \geq e(k) - j\} .$$

¿From Lemma 9 we get the following theorem.

Theorem 16. *Let C be an $[n,k,d]$ code with $e(k) = 1$. Then*
(a) $R_0(C) = k$
or (b) $2 \leq R_0(C) \leq k-1$ *and* $\lceil d/2^{R_0(C)-1} \rceil$ *is odd.*

Next we will show that the necessary condition in Theorem 16 is sufficient in almost all cases.

Theorem 17. *Suppose R is an integer such that $2 \leq R \leq k$ and let m be an integer such that*
(a) $1 \leq m \leq 2^{k-1}$
and (b) *if* $2 \leq R \leq k-1$, *then* $\lceil m/2^{R-1} \rceil$ *is odd,*
then there exists an integer $D(R,m)$ with the property that
if $d \geq D(R,m)$ *and* $d \equiv m \pmod{2^{k-1}}$,
then there exists an $[n,k,d]$ code for which $R_0(C) = R$.

Proof. The proof is by an explicit construction which is a modification of the Solomon-Stiffler construction of codes meeting the Griesmer bound (i.e. with $e(k) = 0$). As before, let S_k denote a $k \times (2^k - 1)$ matrix having all the non-zero vectors in $GF(2)^k$ as columns. For $u < k$ let T_u be a $k \times (2^u - 1)$ matrix of the of the form

$$\begin{bmatrix} S_u \\ 0 \end{bmatrix} ,$$

that is, the upper u rows constitute S_u and the lower $k-u$ rows are all zero.

The integer m has a unique representation of the form

$$m = 2^{k-1} - \sum_{j=1}^{p} 2^{u_j - 1} ,$$

where $p \geq 0$ and $k > u_1 > u_2 > \cdots > u_p \geq 1$. Note that if $R < k$, then $\lceil m/2^{R-1} \rceil$ is odd if and only if $r = u_J$ for some J. Let $s \geq \max\{1, p\}$, and let

$$G = sS_k \setminus \{T_{u_1} \cup T_{u_2} \cup \cdots \cup T_{u_p}\} ,$$

that is, the matrix obtained by removing $T_{u_1}, \cdots, T_{u_p}$ from a matrix whose columns are s copies of S_k. Further, let E_r be the code generated by the first r rows in G_s.

If D is a subspace of the code generated by S_k, then

$$w_S(D) = 2^k - 2^{k - dim(D)} .$$

Hence, if D is a subcode of E_k, then

$$w_S(D) = s(2^k - 2^{k-dim(D)}) - \sum_{j=1}^{p} (2^{u_j} - 2^{u_j - dim(E_{u_j} \cap D)}) .$$

In particular, if $u_J < r \leq u_{J-1}$, then

$$w_S(E_r) = s(2^k - 2^{k-r}) - \sum_{j=1}^{J-1} (2^{u_j} - 2^{u_j - r}) - \sum_{j=J}^{p} (2^{u_j} - 1) = g(r,d) .$$

This shows that $\{E_1, E_2, \cdots, E_k\}$ is a chain for E_k.

Let $r = u_J > 1$ or $r = k$. If $D \neq E_r$ is a subcode of E_k of dimension r, then $dim(E_r \cap D) < r$, and so

$$w_S(D) - w_S(E_r) = \sum_{j=1}^{J-1}(2^{u_j - dim(E_{u_j} \cap D)} - 2^{u_j - r}) + \sum_{j=J}^{p}(2^{u_j - dim(E_{u_j} \cap D)} - 1) \geq 1.$$

Let $G^+ = (G|\bar{x})$ where $x_r = 1$ and $x_i = 0$ for $i \neq r$. Let E_j^+ be the code generated by the j first rows of G^+. Then

$$w_S(E_{r-1}^+) = w_S(E_{r-1}) = g(r-1, d)$$

and

$$w_S(E_r^+) = w_S(E_r) + 1 = g(r, d) + 1 \ .$$

Let $D^+ \neq E_r^+$ be a subcode of E_k^+ of dimension r, and let D be the code obtained by puncturing the last position. Then D is a subcode of E_k of dimension r and $D \neq E_r$. Hence

$$w_S(D^+) \geq w_S(D) \geq w_S(E_r) + 1 = g(r, d) + 1 \ .$$

Therefore, $d_r(E_r^+) = g(r, d) + 1$, and so $R_0(E_k^+) = R$. Since $d_1(E_k^+) = (s - 1)2^{k-1} + m$ we get $D(R, m) \leq \max\{0, p-1\}2^{k-1} + m$. □

As an example, consider $k = 5$. We have determined $D(R, m)$ in all cases as follows: For each R, $2 \leq R \leq 5$ we first determine the possible m as given in the theorem. Then a computer search found the least $d \equiv m \pmod{2^{k-1}}$ for which there exists an $[n, k, d]$ of the required form. By Lemma 13, this d is $D(R, m)$. It turned out that $D(R, m) = m$ in all cases, with one exception, namely $D(2, 2) = 18$.

¿From Lemmas 9 and 10 we get the following theorem.

Theorem 18. *Let C be an $[n, k]$ code with $e(k) = 2$. Then*
(a) $R_0(C) = k$ *and* $R_1(C) \in \{k-1, k\}$,
or (b) $R_0(C) = k$ *and* $R_1(C) < k-1$ *and* $\delta(R_1(C)) \not\equiv 0 \pmod 4$,
or (c) $R_1(C) < R_0(C) < k$ *and* $\delta(R_0(C))$ *is odd and* $\delta(R_1(C)) \not\equiv 0 \pmod 4$.

Combining this with Lemma 13 we get the following.

Theorem 19. *Let C be an $[n, k]$ code with $e(k) = 2$ which satisfies the chain condition. Then*
(a) $R_0(C) = k$ *and* $R_1(C) \in \{k-1, k\}$,
or (b) $R_0(C) = k$ *and* $R_1(C) < k-1$ *and* $\delta_{R_1(C)}$ *is odd,*
or (c) $R_1(C) < R_0(C) < k$ *and* $\delta(R_0(C))$ *is odd and* $\delta(R_1(C))$ *is odd.*

Again the necessary condition in Theorem 19 is sufficient in almost all cases.

Theorem 20. *Let R_0 and R_1 be integers such that $2 \le R_1 \le R_0 \le k$. Let m be an integer such that $1 \le m \le 2^{k-1}$. If*
(a) *$R_0 = k$ and $R_1 \in \{k-1, k\}$,*
or (b) *$R_0 = k$ and $R_1 < k-1$ and $\lceil m/2^{R_1-1} \rceil$ is odd,*
or (c) *$R_1 < R_0 < k$ and $\lceil m/2^{R_0-1} \rceil$ is odd and $\lceil m/2^{R_1-1} \rceil$ is odd,*
then there exists an integer $D(R_0, R_1, m)$ with the property that
if $d \ge D(R_0, R_1, m)$ and $d \equiv m \pmod{2^{k-1}}$,
then there exists an $[n, k, d]$ code with $R_0(C) = R_0$ and $R_1(C) = R_1$ which satisfies the chain condition.

Proof. The proof is similar to the proof of Theorem 17. The only difference is that we now add *two* columns $\bar{x}, \bar{y}$ to G, where $x_{R_0} = 1$, $x_i = 0$ for $i \ne R_0$ and $y_{R_1} = 1$, $y_i = 0$ for $i \ne R_1$. We skip the details. □

We again use $k = 5$ as an illustration. We consider $D(R_0, R_1, m)$ as defined in Theorem 20 and $D'(R_0, R_1, m)$ which is the least $d \equiv m \pmod{2^{k-1}}$ for which there exists an $[n, k, d]$ code (satisfying the chain condition or not). For some combinations of (R_0, R_1, m) there are no $[n, k, d]$ codes with $d \equiv m \pmod{2^{k-1}}$, for any d, by Theorem 18 or Lemma 6. For all other combinations of (R_0, R_1, δ) it turns out that $D'(R_0, R_1, m) = m$. Moreover, it turns out that if there exist $[n, k, d]$ codes with $d \equiv m \pmod{2^{k-1}}$ (that is, if $D(R_0, R_1, m)$ is defined, then $D(R_0, R_1, m) = D'(R_0, R_1, m) = m$ in all cases. We list the various possibilities in the following table.

		excess	m															
R_0	R_1	sequence	1	2	3	4	5	6	7	8	9	10	11	12	13	14	15	16
5	5	0 0 0 0 2	+	+	+	+	+	+	+	+	+	+	+	+	+	+	+	+
5	4	0 0 0 1 2	+	+	+	+	+	+	+	+	+	+	+	+	+	+	+	+
5	3	0 0 1 1 2	+	+	+	+	×	×	×	b	+	+	+	+	a	a	a	a
5	2	0 1 1 1 2	+	+	×	b	+	+	a	a	+	+	×	b	+	+	a	a
4	3	0 0 1 2 2	+	+	+	+	+	+	+	+	a	a	a	a	a	a	a	a
4	2	0 1 1 2 2	+	+	×	b	+	+	a	a	a	a	a	a	a	a	a	a
3	2	0 1 2 2 2	+	+	+	+	a	a	a	a	+	+	+	+	a	a	a	a

Notations used in the table:
+ $D(R_0, R_1, m) = m$,
× $D'(R_0, R_1, m) = m$, codes satisfying the chain condition does not exist by Lemma 12,
a there are no such codes by Theorem 18,
b there are no such codes by Lemma 6 (a).

In particular, for $k = 5$ the necessary conditions of Theorem 18 are also sufficient. Whether this is true for general k and almost all d is an open question.

It is possible to get results similar to Theorem 18 for codes with $e(k) = 3$, $e(k) = 4$, etc., but they are getting increasingly complicated to state (and prove).

References

1. H. Chung, "The second generalized Hamming weight of double-error correcting BCH codes and their dual codes", in: *Algebraic Algorithms and Error-Correcting Codes Conf.*, New Orleans, LA, 1991.
2. H. Chung, "The second generalized Hamming weight of double-error correcting BCH codes and their dual codes", in: *Lecture Notes in Comp. Science*, vol. 539, 118-129, Springer-Verlag 1991.
3. G. Cohen, S. Litsyn, and G. Zémor, "Upperbounds on generalized distances", submitted for publication.
4. G. Cohen, L. Huguet, and G. Zémor, "Bounds on generalized weights", presented at the first French-Israeli Workshop on Algebratic Coding, Paris, July 1993.
5. I. Duursma, H. Stichtenoth, and C. Voß, "Generalized Hamming weights for duals of BCH codes and maximal algebraic function fields", preprint 1993.
6. S. Encheva and T. Kløve, "Codes satisfying the chain condition", *Reports in informatics*, no. 70, Department of Informatics, University of Bergen, October 1992; to appear in *IEEE Trans. Info. Theory*, vol. IT-40, January 1994.
7. G. L. Feng, K. K. Tzeng, and V. K. Wei, "On the Generalized Hamming Weights of Several Classes of Cyclic Codes", *IEEE Trans. Info. Theory*, vol. IT-38, pp. 1125-1130, 1992.
8. G. D. Forney, "Dimension/length profiles and trellis complexity of linear block codes", manuscript, October 1993.
9. G. van der Geer and M. van der Vlugt, "On generalized Hamming weights of BCH-codes", Report 92-19, Dept. of Mathematics and Computer Science, Univ. of Amsterdam, 1992.
10. G. van der Geer and M. van der Vlugt, "The second generalized Hamming weight of the dual codes of double-error correcting binary BCH-codes", Manuscript, December 1992.
11. J. H. Griesmer, "A bound for error-correcting codes", *IBM J. Res. Develop.*, vol. 4, 532-542, 1960.
12. T. Helleseth, T. Kløve, J. Mykkeltveit, "The weight distribution of irreducible cyclic codes with block lengths $n_1((q^l - 1)/N)$", *Discrete Math.*, vol. 18, pp. 179-211, 1977.
13. T. Helleseth, T. Kløve, Ø. Ytrehus, "Generalized Hamming weights of linear codes", *IEEE Trans. Inform. Theory*, vol. IT-38, pp. 1133-1140, 1992.
14. T. Helleseth, T. Kløve, Ø. Ytrehus, "Codes and the chain condition", *Proc. Int. workshop on Algebraic and Combinatorial Coding Theory*, Voneshta Voda, Bulgaria June 22-28, 1992, pp. 88-91.
15. T. Helleseth, T. Kløve, Ø. Ytrehus, "Excess sequences of codes and the chain condition", *Reports in informatics*, no. 65, Department of Informatics, University of Bergen, July 1992.
16. T. Helleseth, T. Kløve, Ø. Ytrehus, "Codes, weight hierarchies, and chains", *Proc. ICCS/ISITA '92*, Singapore, 16-20 November 1992, 608-612.
17. T. Helleseth, T, Kløve, V. Levenshtein, and Ø. Ytrehus, "Bounds on the minimum support weights", submitted for publication.
18. T. Helleseth and P. V. Kumar, "On the weight hierarchy of semiprimitive codes", submitted for publication.
19. T. Helleseth and P. V. Kumar, "The weight hierarchy of the Kasami codes", to appear in *Discrete Math.*.

20. J. W. Hirschfeld, M. A. Tsfasman, and S. G. Vladut, "The weight hierarchy of higher-dimensional Hermitian codes", manuscript, August 1992.
21. G. Kabatianski, "On second generalized Hamming weight", *Proc. Int. workshop on Algebraic and Combinatorial Coding Theory*, Voneshta Voda, Bulgaria June 22-28, 1992, pp. 98-100.
22. T. Kløve, "The weight distribution of linear codes over $GF(q^l)$ having generator matrix over $GF(q)$, *Discrete Mathematics*, vol. 23, pp. 159-168, 1978.
23. T. Kløve, "Support weight distribution of linear codes", *Discrete Math.*, vol. 106/107, pp. 311-316, 1992.
24. T. Kløve, "Minimum support weights of binary codes", *IEEE Trans. Inf. Theory*, vol. IT-39, pp. 648-654, 1993.
25. D. Nogin, "Generalized Hamming weights for codes on multi-dimensional quadrics", *Problems Inform. Transmission*, to appear.
26. J. Simonis, "The effective length of subcodes", manuscript, 1992.
27. H. Stichtenoth and Conny Voß, "Generalized Hamming weights of trace codes", manuscript, 1993.
28. M. van der Vlugt, "On generalized Hamming weights of Melas codes", Report, Mathematical Inst., Univ. of Leiden, October 1992.
29. Zhe-xian Wan, "The weight hierarchies of the projective codes from nondegenerate quadrics", manuscript, 1992.
30. V. K. Wei, "Generalized Hamming Weights for Linear Codes", *IEEE Trans. Inform. Theory*, vol. IT-37, pp. 1412-1418, 1991.
31. V. K. Wei and K. Yang, "On the generalized Hamming weights of squaring construction and product codes", manuscript, 1991.
32. K. Yang, P. V. Kumar, and H. Stichtenoth, "On the weight hierarchy of geometric Goppa codes", preprint, 1992.

Codes that Correct Two-Dimensional Burst Errors

Ernst M. Gabidulin and Vitaly V. Zanin

Moscow Institute of Physics and Technology
Institutskii per. 9
141700 Dolgoprudnyi RUSSIA
e-mail: gab@re.mipt.su

Abstract. Metrics are defined which describe array bursts. Spectra for these metrics, bounds for optimal codes and code constructions are discussed.

1 Introduction

There are very many ways to store and to transmit information. The properties of the channel the information is transmitted through and the properties of errors occurring in the received word are closely connected. In many cases two-dimensional channels are used to store and to transmit information. Examples include different memory chips, floppy disks or magnetic tape. Very frequently the most probable error pattern is a spot. It is easy to describe such errors in terms of two-dimensional bursts.

To describe the features of the channel we introduce special metrics connected with the channel. It is a two-dimensional burst metric. This metric is a particular case of the so-called combinatorial metric defined by E. M. Gabidulin. In papers [11], [12], [13] he developed the theory of combinatorial metrics and the codes defined over them.

For several cases we found the spectrum of the metric, and derived upper and lower bounds on codes over this metric.

There are two ways to construct codes over this metric. One of them is to combine codes over different metrics and the other is to construct codes in terms of their parity check matrices. Imai [1], [2] had constructed two-dimensional burst correcting codes generalizing one-dimensional Fire codes [3]. K.A.S. Abdel-Ghafar, R.J. McEliece and H.C.A. van Tilborg constructed a class of Two-Dimensional Burst Identification and Location codes correcting two-dimensional bursts [4]. Many constructions correcting two-dimensional phased bursts were proposed by P.J. Farrell, R.J. McEliece, M. Sayano, R.M. Goodman, M. Blaum, R.M. Roth, T. Fuja, C. Heegard, H.C.A. van Tilborg, W. Zhang and J.K. Wolf [5], [6], [7], [8], [9], [10].

2 Two-Dimensional Burst Metric and its Properties

Let us consider a space V that consists of binary $n_1 \times n_2$ matrices where n_1 is the number of rows and n_2 is the number of columns. It is a linear space and its dimension is equal to $n_1 n_2$. The linear code $C(n_1 \times n_2, k)$ is a k-dimensional subspace of space V. We define for each matrix A its burst norm $w_{b_1 b_2}(A)$ as the minimum number of the submatrices of size $b_1 \times b_2$ to cover all non-zero elements of the matrix A. It was shown [14] that the function $w(A)$ complies with the definition of metric.

The spectrum of the space V is $\Re^{b_1 b_2}_{n_1 n_2}(d)$–the number of $n_1 \times n_2$ matrices of $b_1 \times b_2$-burst norm d.

Any single burst in a matrix can be described in terms of its pattern and starting position. We say that (u_1, u_2) is the starting position of the burst A if the $b_1 \times b_2$-submatrix B with its right upper corner being at the position (u_1, u_2) has non-zero first row and column and covers all non-zero elements of matrix A. The element distribution pattern of submatrix B is then the pattern of the burst. A cyclic burst is an arbitrary cyclic shift of the submatrix B in both vertical and horizontal dimensions of the matrix. For example, the matrix with non-zero elements in all four corners is a shift by n_1 in vertical and by n_2 in horizontal direction of 2×2-burst situated at the right upper corner of the matrix. It was shown in [4] that the necessary and sufficient condition for all $b_1 \times b_2$-cyclic bursts $[a_{i,j}], 0 \le i < n_1, 0 \le j < n_2$, to have unique patterns and starting positions is

$$n_1 \ge 2b_1 - 1,\ n_2 > 2b_2 - 1.$$

We assume that this condition is true.

It was proved in [4] that the number of distinct $b_1 \times b_2$ burst patterns $N(b_1, b_2)$ is given by the equation:

$$N(b_1, b_2) = 2^{b_1 b_2 - 1} + (2^{b_1 - 1} - 1)(2^{b_2 - 1} - 1) \times 2^{(b_1 - 1)(b_2 - 1)}$$

It can be easily shown that the number of matrices with $d = 1$ for cyclic bursts is as follows:

$$\Re^{b_1 b_2}_{n_1 n_2}(1) = n_1 n_2 N(b_2, b_2)$$

We obtained the spectrum for some specific cases but it proved to be too complicated to do it in the general case for weights greater than 1. In [11] the formula was derived for $n_1 = b_1$ using a recurring equation:

$$\Re^{b_1 b_2}_{n_1 n_2}(d) = \Re^{b_1 b_2}_{n_1 (n_2 - 1)}(d) 2^{b_1 (b_2 - 1)}(2^{b_2} - 1)\Re^{b_1 b_2}_{n_1 (n_2 - b_2)}(d - 1).$$

Using the generating function

$$F = \textstyle\sum_{n_2=0}^{\infty} \Re^{b_1 b_2}_{n_1 n_2}(d) z^n$$

$$= \left[2^{b_1 (b_2 - 1)}(2^{b_2} - 1) z^{b_2}\right]^{d-1} \times \textstyle\sum_{n_2=0}^{\infty} (\sum_{j=0}^{n} \binom{j+d-2}{d-2} \Re^{b_1 b_2}_{n_1 (n_2 - j)}(1)) z^n,$$

where

$$\Re_{n_1 n_2}^{b_1 b_2}(1) = \begin{cases} 2^{b_1 n_2} - 1, \\ \quad if\ 0 < n_2 \leq b_2; \\ 2^{b_1(b_2-1)}(2^{b_2}-1)(n_2 - b_2 + 1) \\ \quad +2^{b_1(b_2-1)} - 1,\ if\ n_2 > b_2 \end{cases}$$

we can obtain:

$$\Re_{n_1 n_2}^{b_1 b_2}(d) = \begin{cases} 0, \quad if\ n_2 \leq b_2(d-1); \\ \left[2^{b_1(b_2-1)}\left(2^{b_1}-1\right)\right]^{d-1} \times \\ \times \sum_{j=0}^{n_2} \binom{j+d-2}{d-2} \Re_{n_1(n_2-b_2(d-1)-j)}^{b_1 b_2}(1), \\ \quad if\ n_2 > b_2(d-1). \end{cases}$$

When $n_1 = b_1 + 1$, $b_2 = 2$ the recurring equations are too complicated, so we only write here the two-dimensional generating function:

$$\Phi(x,y) = \sum_{d=0}^{\infty} \sum_{n_2=0}^{\infty} \Re_{n_1 n_2}^{b_1 b_2}(d) x^{n_2} y^d =$$

$$= \frac{1+(2^{n_1-2}-1)xy+2^{n_1-2}xy^2-2^{n_1-2}(2^{n_1-2}-1)x^2y^2-2^{2n_1-3}x^2y^3}{1-x-2^{n_1-1}xy-2^{n_1-2}(7\cdot 2^{n_1-2}-5)x^2y-4^{n_1-1}x^2y^2+4^{n_1-1}(2^{n_1-2}-1)x^3y^3}$$

This function allows us to find the asymptotic value of the spectrum using the saddle point method:

$$\Re_{n_1 n_2}^{b_1 b_2}(d) = \frac{1}{(2\pi i)^2} \oint_{c_x} \oint_{c_y} \frac{\Phi(x,y)}{x^{n_2+1} y^{d+1}}\, dx\, dy$$

$$\Re_{n_1 n_2}^{b_1 b_2}(d) \sim \tfrac{1}{2\pi} \sqrt{-\tfrac{2\pi}{n_2 h''(n_2)}}\, \varphi(n_2) e^{n_2 h(\alpha, n_2)},$$
$$n \to \infty, \alpha = \tfrac{d}{n_2} = const, n_1 = b_1 + 1,$$

where $h(\alpha, n_2)$ is a function with maximum in the interval $\alpha \in (0;1)$.

It is clear that the most probable weight of the channel corresponds to the value of α where $h(\alpha, n_2)$ takes its maximum. For bursts of size 2×2 this average weight is about $0.7 n_2$.

Consider the set of matrices with all non-zero elements being inside d different submatrices of size $b_1 \times b_2$. It follows that all matrices of weight d are included in this set. So it is followed by:

$$\Re_{n_1 n_2}^{b_1 b_2}(d) \leq [N(b_1, b_2)]^d \binom{n_1 n_2}{d}$$

The matrices of weight d with all the bursts with fixed starting positions (so called phased bursts) are part of all matrices of weight d. Then the following is correct:

$$\Re_{n_1 n_2}^{b_1 b_2}(d) \geq [N(b_1, b_2)]^d \binom{D_{\max}}{d},$$
$$D_{\max} = \tfrac{n_1 n_2}{b_1 b_2}.$$

For the case of $n_1 = b_1 + 1$ it is possible to get a more precise evaluation of the spectrum of the $b_1 \times b_2$-metric.

The recurring equations in this case are the following:

$$\Re^{b_1 b_2}_{n_1 n_2}(d) = \Re^{b_1 b_2}_{n_1(n_2-1)}(d) + N_1(b_1, b_2)\tilde{\Re}^{b_1 b_2}_{n_1(n_2-1)}(d-1) + 2^{b_1 b_2 + b_2 - 2}\Re^{b_1 b_2}_{n_1(n_2-b_2)}(d-2),$$

where $\tilde{\Re}^{b_1 b_2}_{n_1(n_2-1)}(d-1)$ is the number of matrices with a $b_1 \times (b_2-1)$-block of zeros in the left upper corner of the matrix and

$$N_1(b_1, b_2) = 2^{b_1 b_2}$$

$$+(2^{b_1-1}-1)(2^{b_2-1}-1) \times 2^{(b_1-1)(b_2-1)}$$

It can be shown that the parameter $\tilde{\Re}^{b_1 b_2}_{n_1 n_2}(d)$ satisfies the formula:

$$\tilde{\Re}^{b_1 b_2}_{n_1 n_2}(d) = \alpha \cdot \beta \sum_{i=0}^{b_2-2} \alpha^i \cdot \tilde{\Re}^{b_1 b_2}_{n_1(n_2-b_2+1)}(d-1, i+1)$$

$$+\Re^{b_1 b_2}_{n_1(n_2-b_2+1)}(d-1), \alpha = 2^{b_1-1}, \beta = 2^{b_2-1}$$

where $\tilde{\Re}^{b_1 b_2}_{n_1(n_2-b_2+1)}(d-1, i+1)$ is the number of matrices $[a_{i,j}]$ of corresponding weight and size such that one more additional condition is valid:

$$a_{b_2-1,i} = 1$$

It is clear that $\tilde{\Re}^{b_1 b_2}_{n_1(n_2-1)}(d-1) \le \Re^{b_1 b_2}_{n_1(n_2-1)}(d-1)$, so

$$\Re^{b_1 b_2}_{n_1 n_2}(d) \le \Re^{b_1 b_2}_{n_1(n_2-1)}(d) + N_1(b_1, b_2)\Re^{b_1 b_2}_{n_1(n_2-1)}(d-1) + 2^{b_1 b_2 + b_2 - 2}\Re^{b_1 b_2}_{n_1(n_2-b_2)}(d-2)$$

Using this recurring evaluation and the generating function method we have the following:

$$\Re^{b_1 b_2}_{n_1 n_2}(d) \le \{N_1(b_1, b_2)\}^d \sum_{k=0}^{n_2/b_2} \binom{d}{k} \frac{2^k (n_2 - k b_2 + 1)}{\{N_1(b_1, b_2)\}^k},$$

$$b_2 | n_2;\ n_1 = b_1 + 1$$

3 Code Bounds over TDB-Metric

3.1 Upper bounds

Theorem 1. Every code of size $n_1 \times n_2$ with minimum distance d over two-dimensional $b_1 \times b_2$-bursts has rate bounded by:

$$R_d \le 1 - \frac{d-1}{D_{\max}},$$

or in terms of the redundancy of the code:

$$r_d \ge (d-1) b_1 b_2.$$

Proof:

Let us consider the code C with code distance d and size $n_1 \times n_2$. Then we select d $b_1 \times b_2$-bursts that do not intersect in every code matrix. Now each matrix consists of two regions: Θ which consists of those d bursts and Λ, the rest of the matrix. Then we split the code into the set of subcodes such that in each matrix of the subcode the region Λ is the same. The number of these subcodes is as follows:

$$N_{sc} \leq 2^{n_1 n_2 - d b_1 b_2}$$

The code distance of every subcode is greater or equal to d. Let us change the subcodes in the following way: we add an arbitrary matrix from this very subcode to every subcode matrix. Then we have a new subcode with code distance d. This subcode includes the zero matrix; hence, the minimum weight of the subcode is also d. It follows from the construction of subcodes that the matrices from those subcodes contain only zero elements in the region Λ. Hence, the maximum weight of every subcode is equal to d. If from the other side the minimum weight is d as well, then the weight of each code matrix is equal to d or 0.

So, every subcode consists of matrices with weight d and all the non-zero elements of matrices are within d of those $b_1 \times b_2$-bursts. As the code distance is d, the corresponding burst patterns (with the same position) in different matrices are different. It is followed by the statement that the number of matrices in the subcode is less than $2^{b_1 b_2}$. It is not difficult to show that there exists a code with the above parameters and its cardinality is equal to $2^{b_1 b_2}$.

Summing up all the above we can make the following evaluation:

$$M(C) \leq N_{sc} \times 2^{b_1 b_2} \leq 2^{n_1 n_2 - (d-1) b_1 b_2},$$

where $M(C)$ is the cardinality of the original code C. It is immediately followed by the statement of the theorem.

Now that we know the value of $\Re^{b_1 b_2}_{n_1 n_2}(d)$ we can obtain another upper bound[11]:

$$R_d \leq 1 - \frac{\log_2 \Re^{b_1 b_2}_{n_1 n_2}(\frac{d-1}{2})}{n_1 n_2},$$

Then using the upper evaluation on $\Re^{b_1 b_2}_{n_1 n_2}(d)$ given above, we derive the next upper bound:

$$R_d \leq 1 - \frac{\frac{d-1}{2} \log_2 N(b_1 b_2) + D_{\max} H(\frac{\alpha}{2} - \frac{1}{2D_{\max}})}{n_1 n_2}$$

$$- \frac{\frac{1}{2}(3 + \log_2 D_{\max}(\frac{\alpha}{2} - \frac{1}{2D_{\max}})(1 - \frac{\alpha}{2} + \frac{1}{2D_{\max}}))}{n_1 n_2},$$

where $\alpha = d b_1 b_2 / (n_1 n_2) = d / D_{\max}$, $H(.)$ is the entropy function.

The redundancy of the code then follows:

$$r_d \geq \frac{d-1}{2} \log_2 N(b_1 b_2) + D_{\max} H(\frac{\alpha}{2} - \frac{1}{2D_{\max}})$$
$$- \frac{1}{2}(3 + \log_2 D_{\max}(\frac{\alpha}{2} - \frac{1}{2D_{\max}})(1 - \frac{\alpha}{2} + \frac{1}{2D_{\max}}))$$

In [14] an upper bound was obtained using asymptotical evaluation on the spectrum for 2×2-bursts.

In [4] it was shown that $r_3 \geq 2 b_1 b_2 + \log_2 n_1 n_2 - 2$.

3.2 Lower Bounds

It was shown in [11] that there exist codes over combinatorial metrics such that

$$R_d \geq 1 - \frac{\log_2 \Re^{b_1 b_2}_{n_1 n_2}(d-1)}{n_1 n_2}.$$

For $d = 3$, $b_1 = b_2 = 2$ the lower bound was found using asymptotical evaluation. It comes to zero when $d = 0.7n_2$.

Another lower bound can be derived from the lower evaluation of the spectrum presented in the previous section:

$$R \geq 1 - \frac{(d-1)\log_2 N(b_1 b_2) + n_1 n_2 H(\hat{\alpha} - \frac{1}{n_1 n_2})}{n_1 n_2}$$

$$- \frac{\frac{1}{2}(\log_2 2\pi + \log_2 n_1 n_2(\hat{\alpha} - \frac{1}{n_1 n_2})(1 - \hat{\alpha} + \frac{1}{n_1 n_2})}{n_1 n_2}$$

where $\hat{\alpha} = \alpha \cdot b_1 \cdot b_2$, or in terms of redundancy:

$$\begin{aligned} r_d \leq (d-1)\log_2 N(b_1 b_2) + n_1 n_2 H(\hat{\alpha} - \tfrac{1}{n_1 n_2}) \\ -\tfrac{1}{2}(\log_2 2\pi n_1 n_2 (\hat{\alpha} - \tfrac{1}{n_1 n_2})(1 - \hat{\alpha} + \tfrac{1}{n_1 n_2})) \end{aligned}$$

In [4] it was shown that $r_3 \leq 2b_1 b_2 + \log_2 n_1 n_2$.

4 Codes over TDB-Metric

In this section we shall describe known constructions of two-dimensional burst correcting codes and present our new codes.

4.1 Codes with $d = 2$

These codes can detect any two-dimensional burst.

Theorem 2. For every integer n_1, n_2, b_1, b_2 such that $0 < b_1 < n_1, 0 < b_2 < n_2$ there exists a code with $d = 2$ and $r = b_1 b_2$.

It consists of matrices of the following type:

$$\begin{bmatrix} x^0_0 & x^0_{0,1} \cdots & x^0_{0,b_2-1} & x^1_0 & x^1_{0,1} \\ x^0_{1,0} & x^0_{1,1} \cdots & \cdots & x^1_{1,0} & \cdots \\ \vdots & \vdots & \vdots & \vdots & \vdots \\ x^0_{b_1-1,0} & \cdots \quad \cdots & x^0_{b_1-1,b_2-1} & x^1_{b_1-1,0} & \cdots \\ \vdots & \vdots & \vdots & \vdots & \vdots \end{bmatrix}$$

where

$$\textstyle\sum_{k=0}^{D_{max}} x^k_{i,j} = 0,\ mod2, \quad 0 \leq i, j \leq b_1, b_2,$$

This is simply the interleaving of $b_1 b_2$ codes detecting single Hamming errors. It is not difficult to show that it is an optimal code, i.e. its redundancy is minimal.

4.2 Codes with $d = 3$

There are several constructions of codes correcting one two-dimensional burst. Here we describe and compare some of them.

Theorem 3. For all r, s such that $r|s$ there exists a code C_H of size $n_2 \times n_2$:

$$n_1 = b_1 \left(2^r - 1\right), \quad n_2 = b_2 \tfrac{2^s-1}{2^r-1}$$

and of rate

$$R_H = 1 - \frac{s}{2^s - 1}.$$

It is also the interleaving of $b_1 b_2$ Hamming codes correcting 1-fold errors. Its redundancy is

$$r_H = b_1 b_2 s = b_1 b_2 \log_2 n_1 n_2$$

Theorem 4. For all r, s such that $r|s$, $s \geq 2^r + 1$ there exists a code C_F of size $n_1 \times n_2$ where

$$n_1 = 2^r - 1;$$

$$n_2 = \frac{2^s - 1}{2^r - 1}(2^{r+1} - 1)$$

with code words of the type:

$$\begin{bmatrix} x_0 & x_{n_1} & \cdots & x_{n_1(n_2-1)} \\ x_1 & x_{n_1+1} & \cdots & \cdots \\ x_2 & \cdots & \cdots & \cdots \\ \vdots & \vdots & \vdots & \vdots \\ x_{n_1-1} & x_{2n_1-1} & \cdots & x_{n_1 n_2 - 1} \end{bmatrix},$$

where $(x_0 \ldots x_{n_1 n_2 - 1})$ is a code word from the Fire code [3] correcting one-dimensional bursts of length $n_1 + 2$.

This code can correct any two-dimensional burst of size 2×2 and many other error configurations. The rate of the code is given by:

$$R_F = 1 - \frac{s + 2n_1 + 3}{(2^s - 1)(2n_1 + 3)}$$

The same construction can be used to correct any burst of size $b_1 \times 2$ but with the Fire code correcting $(n_1 + b_1)$-bursts. The rate of the code then is as follows:

$$R_F = 1 - \frac{s + 2(n_1 + b_1) - 1}{(2^s - 1)(2(n_1 + b_1) - 1)}$$

The following table compares codes obtained using the above construction for $d = 3$.

It is clear that the Fire construction is faster but it has a size limitation: the length of the code matrix is exponentially greater than its height.

Table 1.

n_1	s	Code	Code size	r	$1-R$
3		C_H	4092	40	9.78×10^{-3}
	9	C_F	4599	18	3.90×10^{-3}
7		C_H	8188	44	5.37×10^{-3}
	9	C_F	8687	26	2.99×10^{-3}

The construction presented in [1] is a generalization of one-dimensional Fire codes. It is described in terms of the generating matrix of the code. The main idea is to divide the error correction into two parts : 1) to recognize the burst pattern, and 2) to define its starting position. The TDF-code is a cyclic code ..

The BIL-codes described in [4] uses the same idea of separation of the identification of the burst pattern and finding its position. But different ways to implement each of the tasks were presented in the article.

To compare all of these constructions let us use the concept of the excess redundancy described in [4] but slightly changed:

Consider an infinite class $\Im$ of $b_1 \times b_2$-burst correcting codes, and suppose that for every positive integer S, the subset $\Im(S)$ is nonempty. For each $C \subset \Im$, let $n_{1C} \times n_{2C}$ and r_C denote the area and redundancy of C respectively, then we define the *excess redundancy* of the class $\Im$ as

$$\tilde{r}_{\Im}(b_1, b_2) = \lim_{S\to\infty} \inf_{C\subset\Im(S)}(r_C - \log_2(n_1 n_2 N(b_1, b_2))$$

Single burst correction requires all syndromes of single bursts to be different, and thus r_C should be at least $\log_2(n_1 n_2 N(b_1, b_2))$, as the number of different single bursts is equal to $n_1 n_2 N(b_1, b_2)$.

For Interleaving of Hamming Codes the excess redundancy is infinite.

For Construction with One-Dimensional Fire Code (LF) the excess redundancy for fixed n_1 is equal to

$$\tilde{r}_{1\times F}(b_1, b_2 = 2) = 2(n_1 + b_1) - 1 - \log_2(2(n_1 + b_1) - 1) - \log_2 N(b_1, b_2)$$

For Two-Dimensional Fire Code:

$$\tilde{r}_{2\times F}(b_1, b_2) = (2b_1 - 1)(2b_2 - 1) - \log_2(2b_1 - 1)(2b_2 - 1) - \log_2 N(b_1, b_2)$$

For BIL codes it is proved in [4] that

$$2b_1 b_2 - 2 - \log_2 N(b_1, b_2) \le \tilde{r}_{\Im}(b_1, b_2) \le 2b_1 b_2 - \log_2 N(b_1, b_2)$$

For 2×2-bursts $N(b_1, b_2) = 10$, [4]

$$\tilde{r}_{BIL}(2,2) = 7 - \log_2 10$$

$$\tilde{r}_{2\times F}(2,2) = 9 - \log_2 9 - \log_2 10 \simeq 5.8 - \log_2 10$$

$$\tilde{r}_{1\times F}(2,2) = 9 - \log_2 9 - \log_2 10 \simeq 5.8 - \log_2 10, n_1 = 3$$

$$\tilde{r}_{1\times F}(2,2) = 17 - \log_2 17 - \log_2 10 \simeq 12,9 - \log_2 10, n_1 = 7$$

Codes described in Theorem 4 are better in some cases. For example, there exists such a code of size $n_1 \times n_2 = 567$ and redundancy 15 bits, in comparison with the TDF-code of size 180 and redundancy 16 bits.

For $(n_1 - 1) \times 2$ bursts the TDF and BIL codes can only correct the errors if $n_1 = 3$ and for this case the excess redundancy is the following

$$4b_1 - 2 - \log_2 N(b_1, 2) \leq \tilde{r}_{BIL}(b_1, 2) \leq 4b_1 - \log_2 N(b_1, 2)$$

$$\tilde{r}_{2\times F}(b_1, 2) = 3(2b_1 - 1) - \log_2 3(2b_1 - 1) - \log_2 N(b_1, 2)$$

$$\tilde{r}_{1\times F}(2,2) = (4b_1 + 1) - \log_2(4b_1 + 1) - \log_2 N(b_1, 2)$$

The excess redundancy of our codes is the least one.

The following table compares some codes from the above constructions. It shows that for "narrow" codes the LF-construction (Theorem 4) is better.

Table 2. Codes comparison

Code	$n_1 \times n_2$	r	$1 - R$
TDF	$12 \times 15 = 180$	16	8.89×10^{-2}
LF	$3 \times 189 = 567$	15	2.64×10^{-2}
TDF	$21 \times 32 = 672$	18	2.68×10^{-2}
TDF	$3 \times 255 = 765$	17	2.22×10^{-2}
LF	$3 \times 1533 = 4599$	18	3.90×10^{-3}
TDF	$3 \times 1364 = 4092$	22	5.37×10^{-3}
TDF	$12 \times 340 = 4080$	24	5.80×10^{-3}
BIL	$63 \times 63 = 3969$	19	4.7×10^{-3}
LF	$7 \times 9945 = 69615$	29	4.17×10^{-4}
TDF	$20 \times 3276 = 65520$	28	4.27×10^{-4}
BIL	$255 \times 255 = 65025$	23	3.54×10^{-4}

4.3 Codes Correcting Several Bursts

To correct several bursts the best that can be suggested now is to interleave a corresponding quantity of codes correcting several Hamming errors. The speed of the construction is lower than one could expect but it is rather simple in realization. As an example of optimal construction the repetition code described in [14] can serve. Its redundancy is equal to $(D_{\max}-1)b_1b_2$ and its code distance is equal to $D_{\max}$.

References

1. Hideki Imai, "Two-dimensional Fire codes," *IEEE Trans. Inform. Theory, vol. IT-19, #6, Nov.. 1973..*
2. Hideki Imai, "Two-dimensional burst-correcting codes" (in Japanese), *IECE Trans.,vol. 55-A, Aug. 1972.*
3. P. Fire, "A class of multiple error correcting binary codes for non independent errors", *Sylvania Reconnaissance System Lab., Mountain View, Calif., Sylvania Rep. RSL-e-2, 1959.*
4. K.A.S. Abdel-Ghafar, R.J. McEliece and H.C.A. van Tilborg, "Two- dimensional burst identification codes and their use in burst correction", *IEEE Trans. Inform. Theory, vol. IT-3, #3, May 1988.*
5. T. Fuja, M. Blaum, C. Heegard, "Cross parity check convolutional codes," *IEEE Trans. Inform. Theory, vol. IT-35, #6, Nov. 1989.*
6. M. Blaum, R.M. Roth, "New array codes for multiple phased burst correction", *IEEE Trans. Inform. Theory, vol. IT-39, #1, Jan. 1993.*
7. P.J. Farrell, H.C.A. van Tilborg, M. Blaum, "Multiple burst-correcting array codes," *IEEE Trans. Inform. Theory, vol. IT-34, #5, Sep. 1993.*
8. R.J. McEliece, M. Sayano, R.M. Goodman, "Phased burst error-correcting codes," *IEEE Trans. Inform. Theory, vol. IT-39, #2, Mar. 1993.*
9. K.A.S. Abdel-Ghafar, R.J. McEliece, A.M. Odlyzko and H.C.A. van Tilborg, "On the existence of optimum cyclic burst-correcting codes," *IEEE Trans. Inform. Theory, vol. IT-32, #6, Nov. 1986.*
10. W. Zhang and J.K. Wolf, "A class of binary burst correcting quasi-cyclic codes," *IEEE Trans. Inform. Theory ,vol. IT-34, #3, May. 1988.*
11. E.M. Gabidulin, "Combinatirial metrics in coding theory", *2-nd Int.Symposium on Inf. Theory, 1971, pp. 169-176.*
12. E.M. Gabidulin, V.I. Korzhik, Codes correcting Lattice-Pattern Errors, (in Russian), *Izvestya VUZ, Radioelektronika, 1972, vol.15, #4.*
13. E.M. Gabidulin, "Optimal array error correcting codes", (in Russian), *Probl. Peredach. Inform., vol.21., #2, Apr.-June, 1985.*
14. E.M. Gabidulin, V.V. Zanin, "Matrix codes correcting array errors of size 2*2," *International Symposium on Communication Theory & Applications, Proceedings, pp. 207-212, 1993.*

Self-Checking Decoding Algorithm for Reed-Solomon Codes

I.M.Boyarinov

Institute for Problems of Cybernetics
37, Vavilova str., B-312, Moskow, 117312, Russia

Abstract. The definitions of error secure and self-checking algorithms are given. It is shown that decoding algorithms for Reed-Solomon codes over $GF(q)$ can be transformed in self-checking algorithms for a set F of single algorithmic errors. The two kinds of check functions of error secure decoding algorithms of Reed-Solomon codes are described.

1 Introduction

Decoders for error-correcting codes are central to most high-speed communications systems, for the reliable operation of such devices is critical to the continuous flow of information. Such decoders are being implemented in very large scale integration (VLSI), where increased circuit density makes such designs more susceptible to a class of errors called soft errors. In addition, protection against such errors also covers the more classical errors such as the stuck-at types.

Different techniques can be used in fault-tolerant decoders for error-correcting codes [1, 2, 3]. For example, the Berlekamp-Massey algorithm for BCH codes can be protected through a chord recursion property affiliated with the error connection polynomial [1].

In this paper we will introduce the concepts of error secure and self-checking algorithms and apply them to decoding algorithms for Reed-Solomon (RS) codes.

2 Error Secure and Self-Checking Algorithm

We consider an algorithm L that consists of m $(m \geq 1)$ subalgorithms (steps). A subalgorithm L_i $(1 \leq i \leq m)$ consists of a set R_i of elementary operations. For example, an arithmetic operation in $GF(q)$ is an elementary operation of a decoding algorithm of a code over $GF(q)$.

A single error in an algorithm L is the incorrect result of an elementary operation. Denote by F_i a single error set of a subalgorithm L_i. Then $F = F_1 \cup F_2 \cup \ldots \cup F_m$ is a single error set of an algorithm L.

An algorithm L produces an output vector $y(x, f)$, which is a function of an input vector $x \in X$ and an algorithmic error $f \in F$. In the absence of algorithmic errors, an input x from X produces an output $y(x, \lambda)$ from Y (λ is the null error).

A subalgorithm L_i produces a vector $y_i(x_i, f_i)$, which is a function of an input vector x_i of L_i and an algorithmic error $f_i \in F_i$.

An algorithm L is error secure for an input set X and an algorithmic error set F if for any input x in X and any error f in F, either $y(x,f) = y(x,\lambda)$ or $y(x,f) \notin Y$.

A subalgorithm L_i is error secure for an input set X_i and an error set F_i if for any input $x_i \in X_i$ and any error $f_i \in F_i$, either $y_i(x_i,f_i) = y_i(x_i,\lambda)$ or $y_i(x_i,f_i) \notin Y_i$.

If each subalgorithm $L_i (1 \leq i \leq m)$ is error secure for an error set F_i, then an algorithm L is error secure for an error set $F = F_1 \cup F_2 \cup \ldots \cup F_m$.

A check function ψ_i is constructed to observe the outputs of a subalgorithm L_i. By definition the check function

$$\psi_i = \begin{cases} 1 \text{ , if } y(x_i,f_i) \notin Y_i \\ 0 \text{ , otherwise.} \end{cases}$$

If ψ_i $(1 \leq i \leq m)$ is a check function of an error secure subalgorithm L_i, then $\psi = \psi_1 \vee \psi_2 \vee \ldots \vee \psi_m$ is a check function of an error secure algorithm L. The sign $\vee$ denotes the OR function.

A check function, which observes outputs of an error secure algorithm L, is called a code check function.

A self-checking algorithm L, ψ consists of an error secure algorithm L and a check function ψ.

An output of a self-checking algorithm L, ψ is a pair (y, ψ). A single error may be made calculating either an algorithm L or a check function ψ. If $\psi = 0$, then $y(x,f) = y(x,\lambda)$.

3 Self-Checking Decoding Algorithm for Reed-Solomon Codes

Let V denote a code of minimum distance $d = 2t + 1$. Consider a decoding algorithm A_V of a code V such that a received word is decoded to the codeword closest in terms of the Hamming distance. If an input set X of a decoding algorithm A_V is $X = \{\mathbf{x} : \mathbf{x} = \mathbf{v} + \mathbf{e}, \mathbf{v} \in V$, the weight $w(\mathbf{e}) \leq t\}$, then in the absence of algorithmic errors the algorithm decodes correctly.

Now suppose that there can be errors in a decoding algorithm. The decoding algorithm A_V can be transformed into the error secure algorithm A_V^* for a set F of single algorithmic errors. For this we will calculate the weight $w(\mathbf{y}(\mathbf{x},f) - \mathbf{x})$, where $\mathbf{y}(\mathbf{x},f)$ is an output word of A_V and $f \in F$. Let

$$Y^* = \{\mathbf{y}(\mathbf{x},\lambda) : \mathbf{y}(\mathbf{x},\lambda) \in V, w(\mathbf{y}(\mathbf{x},\lambda) - \mathbf{x}) \leq t\}.$$

Lemma 1. *If Y^* is an output set of a decoding algorithm A_V^* in the absence of algorithmic errors, then an algorithm A_V^* is error secure for an input set X and an algorithmic error set F.*

By Lemma 1 an algorithm A_V^*, ψ^* is self-checking if

$$\psi^* = \begin{cases} 1 \text{ , if } \mathbf{y}(\mathbf{x}_i,f_i) \notin Y^* \\ 0 \text{ , otherwise.} \end{cases}$$

For many codes a check function can be constructed simpler than ψ^*. In particular it can be done for Reed-Solomon codes.

There are several techniques for decoding RS codes. The method described here is chosen because of its simplicity [4]. We assume that q is a prime power, and $GF(q)$ denotes the finite field with q elements. Let α be a primitive element of $GF(q)$ and $g(x) = (x-\alpha)(x-\alpha^2)\dots(x-\alpha^{2t})$ be the generator polynomial for the Reed-Solomon code V over $GF(q)$ with the minimum distance $d = 2t+1$.

Suppose that $c(x) = c_0 + c_1x + \dots + c_{n-1}x^{n-1}$ is a code polynomial which is transmitted over a noisy channel, and that $r(x) = r_0 + r_1x + \dots + r_{n-1}x^{n-1}$ is received. Then the error polynomial is defined as $e(x) = r(x) - c(x) = e_0 + e_1x + \dots + e_{n-1}x^{n-1}$. If $e_i \neq 0$, we say that an error has occurred in position α^i.

The decoding algorithm L_{RS} is as follows.

1. Compute the syndromes
$$S_j = \sum_{i=0}^{n-1} r_i\alpha^{(j+1)i} \text{ for } 0 \le j \le 2t-1.$$
Form a syndrome polynomial
$$S(z) = \sum_{j=0}^{2t-1} S_j z^j .$$
2. Using the extended Euclidean algorithm for polynomials z^{2t} and $S(z)$ evaluate the error locator polynomial $\sigma(z)$ and the error evaluator polynomial $\omega(z)$.
3. Find the roots of $\sigma(z)$.
4. Determine the positions and the values of errors.
5. Define the output polynomial $c_{\text{out}}(x) = r(x) - e(x)$.

In the absence of algorithmic errors, $c_{\text{out}}(x) = c(x)$ if the weight $w(e(x)) \le t$.

Now suppose that there can be errors in the decoding algorithm L_{RS}. Furthermore we suppose that $deg\ (\sigma(x)) \le t$ and $deg\ (\omega(x)) \le t-1$ for any error $f \in F'$, where F' is a set of single errors of L_{RS}. It is easy to see that $w(c_{\text{out}}(x)-c(x)) \le 1$ for any single error on the step 4 or the step 5. Therefore the following lemma holds.

Lemma 2. *The weight $w(c_{\text{out}}(x) - c(x)) \le 2t$ for any input polynomial $r(x) = c(x) + e(x), w(e(x)) \le t$ and any error $f \in F'$.*

Lemma 3. *If $(q-1)(q-2)\dots(q-t) > q^{t-1} - q^{t-2}$, there is an algorithmic error $f \in F'$ on the step 2 such that the weight $w(c_{\text{out}}(x) - c(x)) = 2t$ for any input polynomial $r(x) = c(x) + e(x), w(e(x)) \le t$.*

The following theorem is an immediate consequence of lemmas 2 and 3.

Theorem 4. *The decoding algorithm L_{RS} is error secure for the set F' of single algorithmic errors and not error secure for the set F'' of double algorithmic errors.*

If $S_{\text{out}}(z)$ is the syndrome polynomial of the output polynomial $c_{\text{out}}(x)$, then for the set F' of single algorithmic errors the code check function of the algorithm L_{RS} is

$$\psi_i = \begin{cases} 1 \text{ , if } S_{\text{out}}(z) \neq 0 \\ 0 \text{ , otherwise} \end{cases}$$

Each step of the decoding algorithm L_{RS} can be made self-checking. It can be useful for practical reasons.

Introduce a function $u(g), g \in GF(q)$ such that

$$u(g) = \begin{cases} 1 \text{ , if } g \neq 0 \\ 0 \text{ , otherwise} \end{cases}$$

Suppose that all $S_j, j = 0, 1, \ldots, 2t-1$ are calculated independently.

Calculate

$$\beta_i = \sum_{j=0}^{2t-1} \alpha^{(j+1)i}, S^*_{2t} = -\sum_{i=0}^{n-1} r_i\beta_i \text{ and } g_1 = \sum_{j=0}^{2t-1} S_j + S^*_{2t}$$

Then the sequence $S_0, S_1, \ldots, S_{2t-1}, S^*_{2t}$ is the word of the linear code over $GF(q)$ of minimum distance $d = 2$. Now step 1 of the decoding algorithm L_{RS} is self-checking and the check function $\psi_1 = u(g_1)$.

On step 2 the error locator polynomial $\sigma(z)$ and the error evaluator polynomial $\omega(z)$ can be found by the extended Euclidean algorithm for the polynomials z^{2t} and $S(z)$ as follows.

Let

$$s_0(z) = 1, t_0(z) = 0, r_0(z) = z^{2t},$$
$$s_1(z) = 0, t_1(z) = 1, r_1(z) = S(z).$$

For $i \geq 2$,

$$s_i(z) = s_{i-2}(z) - q_i(z)s_{i-1}(z), \tag{1}$$
$$t_i(z) = t_{i-2}(z) - q_i(z)t_{i-1}(z) \tag{2}$$

where $q_i(z)$ is determined by applying the division algorithm to $r_{i-2}(z)$ and $r_{i-1}(z)$,

$$r_{i-2}(z) = q_i(z)r_{i-1}(z) + r_i(z) \tag{3}$$

with $r_i(z) = 0$ or $deg\ r_i(z) < deg\ r_{i-1}(z)$.

Iterations continue until $deg\ r_{i-1}(z) \geq t$ and $deg\ r_i(z) < t$. For the first such i, the polynomials $\sigma(z) = t_i(z)$ and $\omega(z) = r_i(z)$ are the unique solution of the key equation

$$\omega(z) \equiv \sigma(z)S(z) \pmod{z^{2t}}$$

for $\sigma(0) = 1, deg\ \sigma(z) \leq t$ and $deg\ \omega(z) \leq t-1$ [4].

If F_e is a set of single algorithmic errors of the extended Euclidean algorithm, then

$$F_e = F_e^{(1)} \vee F_e^{(2)} \vee F_e^{(3)},$$

where F_e^1 is a set of single arithmetic errors in 1,2 and 3, $F_e^{(2)}$ is a set of errors in comparison operations on $deg\ r_{j-1}(z)$ and $deg\ r_j(z)$ for all $j (j \le i)$, $F_e^{(3)}$ is a set of errors in comparison operations on $deg\ r_j(z)$ and t for all $j (j \le i)$ and i is the first such that $deg\ r_{i-1}(z) \ge t$ and $deg\ r_i(z) < t$. We will use the following properties of the extended Euclidean algorithm.

Lemma 5. *[4] For $i \ge 1$,*

$$s_i(z)z^{2t} + t_i(z)S(z) = r_i(z) \tag{4}$$

and

$$deg\ r_i(z) + deg\ t_{i+1}(z) = 2t. \tag{5}$$

Furthermore we will carry out the step i of the division algorithm (3) only if $deg\ r_{i-2}(z) > deg\ r_{i-1}(z)$. Then the following Lemma holds.

Lemma 6. *The algorithm for calculating $\sigma(z)$ and $\omega(z)$ is self-checking with the check function ψ_2,*

$$\psi_2 = \psi_2^{(1)} \vee \psi_2^{(2)}$$

where

$$\psi_1^{(2)} = \begin{cases} 0, \text{ if} & s_i(z)z^{2t} + t_i(z)S(z) = r_i(z) \text{ for } i \text{ such that} \\ & deg\ r_{i-1}(z) \ge t \text{ and } deg\ r_i(z) \\ 1, \text{ otherwise} & \end{cases}$$

and

$$\psi_2^{(2)} = \begin{cases} 0, \text{ if} & deg\ r_j(z) + deg\ t_{j+1}(z) = 2t \text{ and} \\ & deg\ r_{j-1}(z) > deg\ r_j(z) \text{ for } j, 1 \le j \le i \\ 1, \text{ otherwise.} & \end{cases}$$

On step 3 the roots of $\sigma(z) = t_i(z)$ are found. Suppose that under finding the roots $z_0, z_1, \dots, z_{l-1}$ of $\sigma(z) = \sigma_0 z^l + \sigma_1 z^{l-1} + \dots + \sigma_l (l \le t)$ can be an error f from a set F_3 of single algorithmic errors. Calculate $\sigma'(z) = \sigma_0(z - z_0)(z - z_1)\dots(z - z_{l-1})$. Then step 3 is self-checking with the check function $\psi_3 = u(g_3), g_3 = \sigma(z) - \sigma'(z)$.

It easy to see that any single algorithmic error on step 4 or step 5 can lead to no more than a single error in the output polynomial $c_{out}(x)$. Calculate $g_{4,5} = c_{out}(\alpha)$. Then step 4 and step 5 are self-checking with the check function $\psi_{4,5} = u(g_{4,5})$.

Theorem 7. *The decoding algorithm of the Reed-Solomon code V over $GF(q)$ of the minimum distance $d = 2t + 1$ is self-checking for an input set $X = \{\mathbf{x} : \mathbf{x} = \mathbf{v} + \mathbf{e}, \mathbf{v} \in V, w(\mathbf{e}) \le t\}$ and a set F of single algorithmic errors if the check function is*

$$\psi = \psi_1 \vee \psi_2 \vee \psi_3 \vee \psi_{4,5}.$$

4 Conclusion

The concepts of error secure and self-checking algorithms have been introduced. For a set F of single algorithmic errors self-checking decoding algorithms of Reed-Solomon codes over $GF(q)$ of the minimum distance $d = 2t + 1$ are constructed. The two kinds of check functions of error secure decoding algorithms of Reed-Solomon codes are described.

References

1. G.R.Redinbo, Fault-tolerant decoders for cyclic error- correcting codes. *IEEE Trans. Comput.*, V. C-36. No. 1, Jan.1987, pp. 47 - 63.
2. J.F.Wakerly, Error Detecting Codes, Self-Checking Circuits and Applications. North-Holland, New York, 1978.
3. T.R.N.Rao and E.Fujiwara, Error-control Coding for Computer Systems. Prentice-Hall,. Englewood Cliffs, N.Y.,1989.
4. F.J.MacWilliams and N.J.A.Sloane, The Theory of Error-Correcting Codes. North-Holland, Amsterdam, 1977.

Partial Unit Memory Codes on the Base of Subcodes of Hadamard Codes

V.V.Zyablov, A.E.Ashikhmin

Institute for Problems of Information Transmission
Russian Academy of Sciences
Ermolova 19, Moscow GSP-4, 101447, Russia
E-mail: aea@ippi.msk.su

Abstract. Partial memory codes on the base of subcodes of nonlinear Hadamard codes are proposed. It is shown that they have better parameters than partial memory codes based on the linear codes.

1 Introduction

Unit memory (UM) codes are well known [1]. Here we consider the partial unit memory (PUM) codes and codes with partial memory two [2] based on subcodes of nonlinear Hadamard codes. Although ensembles of nonlinear PUM codes were already considered [3], up to now there is no known example in which nonlinear PUM codes have better performances than linear PUM ones. Here we present two examples of this kind.

2 Constructing of nonlinear partial memory codes

We will construct nonlinear partial memory codes on the base of regular trellis with the following properties. Each node in the i-th column is connected with each node in the (i+1)-th column by branches; each branch is labeled with a binary vector of length n. The number of the branches coming from one node to another node is the same; the sets of such branches we call bunches. The vectors on the branches leaving or coming to some node can be obtained by adding a certain vector to the vectors on the branches leaving or coming to the zero node. Each path from left to right through this trellis specifies a codeword of a nonlinear partial memory code. Note that the above trellis is almost the trellis corresponding to a linear partial memory code. The only difference between them is that in the case of a linear partial memory code, vectors on the branches leaving the zero node and vectors on the branches coming to the zero node are codewords of linear block codes, whereas in the case of a nonlinear partial memory code these vectors are codewords of nonlinear block codes. We will denote nonlinear partial memory codes by $(n, M|M_1)$, where n is the length of the vectors on the branches, M is the number of branches leaving one node and M_1 is the number of nodes in a column. Hence M/M_1 is the number of codewords in a bunch.

We introduce the following notation. Let $Q = \{\mathbf{q}^{(0)}, ..., \mathbf{q}^{(M-1)}\}$ be an (n, M) nonlinear block code of blocklength n and cardinality M and let $\mathbf{q}^{(0)}, \ldots, \mathbf{q}^{(M-1)}$ be its codewords. Let us introduce an operation $\triangleright$ that maps an index $x \in \{0, \ldots, M-1\}$ into the corresponding codeword $\mathbf{q}^{(\mathbf{x})}$ from Q, i.e. $x \triangleright Q = \mathbf{q}^{(\mathbf{x})}$. Let $Q_1, Q_2, \ldots Q_l$ be $(n, M_1), (n, M_2), \ldots, (n, M_l)$ nonlinear block codes respectively. Let us introduce a sum of $Q_1, Q_2, \ldots, Q_l$ as follows:

$$Q = \{\mathbf{q}^{(0)}, ..., \mathbf{q}^{(\mathbf{M_1 M_2 \ldots M_l - 1})}\} \begin{bmatrix} Q_1 \\ Q_2 \\ \vdots \\ Q_l \end{bmatrix} =$$

$$= \{\mathbf{q}_1^{(\mathbf{i})} + \mathbf{q}_2^{(\mathbf{j})} + \ldots + \mathbf{q}_l^{(\mathbf{k})} : i = 0, \ldots, M_1 - 1; j = 0, \ldots, M_2 - 1; \ldots; k = 0, \ldots, M_l - 1\}.$$

Then an operation $\triangleright$ mapping the indexes $x^{(1)} \in \{0, \ldots, M_1-1\}, x^{(2)} \in \{0, \ldots, M_2-1\}, \ldots, x^{(l)} \in \{0, \ldots, M_l - 1\}$ into the codeword from Q is determined in the following way:

$$(x^{(1)}, x^{(2)}, \ldots, x^{(l)}) \triangleright Q = x^{(1)} \triangleright Q_1 + x^{(2)} \triangleright Q_2 + \ldots + x^{(l)} \triangleright Q_l.$$

The sign '+' denotes component-wise modulo-2 addition of vectors. Matrices of the type:

$$\begin{bmatrix} Q_1 \\ Q_2 \\ \vdots \\ Q_l \end{bmatrix}$$

we will call quasi-generator matrices. Then the quasi-generator matrix of a nonlinear PUM code on the base of codes $Q_{00} = (n, M/M_1), Q_{01} = (n, M_1), Q_{11} = (n, M_1)$ is determined as follows:

$$G = \begin{bmatrix} Q_{00} & \\ Q_{01} & Q_{11} \end{bmatrix}. \tag{1}$$

The vectors $v_{i,j,l}$ belonging to the bunch coming from node i to node j are determined as follows:

$$v_{i,j,l} = q_{11}^{(i)} + q_{01}^{(j)} + q_{00}^{(l)};\ i, j = 0, \ldots, M_1 - 1;\ l = 0, \ldots, M/M_1 - 1.$$

An output vector $\mathbf{z_i}$ of length n is calculated from the rule

$$\mathbf{z_i} = (x_i^{(0)}, x_i^{(1)}) \triangleright \begin{bmatrix} Q_{00} \\ Q_{01} \end{bmatrix} + x_{i-1}^{(1)} \triangleright Q_{11}.$$

The quasi-generator matrix of nonlinear code with partial memory two on the base of codes $Q_{00}, Q_{01}, Q_{02}, Q_{11}, Q_{22}$ is determined in a similar manner:

$$G = \begin{bmatrix} Q_{00} & & \\ Q_{01} & Q_{11} & \\ Q_{02} & & Q_{22} \end{bmatrix}. \tag{2}$$

An output vector $\mathbf{z_i}$ of length n is calculated from the rule

$$\mathbf{z_i} = (x_i^{(0)}, x_i^{(1)}, x_i^{(2)}) \triangleright \begin{bmatrix} Q_{00} \\ Q_{01} \\ Q_{02} \end{bmatrix} + x_{i-1}^{(1)} \triangleright Q_{11} + x_{i-2}^{(2)} \triangleright Q_{22}.$$

2.1 Construction 1

Now we will construct a $(12, 24|12)$-nonlinear PUM code with free distance 12 and $\alpha = 2$. Let A be a matrix obtained from a Hadamard matrix of order 12

$$A = \begin{bmatrix}
1&0&1&0&0&0&1&1&1&0&1\\
1&1&0&1&0&0&0&1&1&1&0\\
0&1&1&0&1&0&0&0&1&1&1\\
1&0&1&1&0&1&0&0&0&1&1\\
1&1&0&1&1&0&1&0&0&0&1\\
1&1&1&0&1&1&0&1&0&0&0\\
0&1&1&1&0&1&1&0&1&0&0\\
0&0&1&1&1&0&1&1&0&1&0\\
0&0&0&1&1&1&0&1&1&0&1\\
1&0&0&0&1&1&1&0&1&1&0\\
0&1&0&0&0&1&1&1&0&1&1
\end{bmatrix}.$$

Let $A^* = [a_{11}, a_{10}, ..., a_1]$, where a_i is the i-th column of A. We form two different Hadamard matrices H_{12}, H_{12}^* as follows:

$$H_{12} = \begin{bmatrix} 0 & \mathbf{0} \\ \mathbf{0}^t & A \end{bmatrix}, \quad H_{12}^* = \begin{bmatrix} 0 & \mathbf{0} \\ \mathbf{0}^t & A^* \end{bmatrix}.$$

Then we form the codes Q_{00}, Q_{01}, Q_{11} as follows:

$$Q_{00} = \{(0,0,0,0,0,0,0,0,0,0,0,0), (1,1,1,1,1,1,1,1,1,1,1,1)\},$$

the codewords of Q_{01} are rows of H_{12} and the codewords of Q_{11} are rows of H_{12}^*. It is easy to observe that a nonlinear PUM code with a quasi-generator matrix (1) has free distance equaling 12. The minimum distance of code

$$Q = \begin{bmatrix} Q_{00} \\ Q_{01} \\ Q_{11} \end{bmatrix}$$

equals 2, therefore $\alpha = 2$. Thus, we obtain a $(12, 24|12)$-nonlinear PUM code with $d_{free} = 12$ and $\alpha = 2$. Note that it is impossible to construct this code using linear block codes. Moreover, the $(12, 16|8)$-linear PUM code is also not known.

Now let H_{24}, H_{24}^* be formed from Sylvester's rule

$$H_{24} = \begin{bmatrix} H_{12} & H_{12} \\ H_{12} & -H_{12} \end{bmatrix}, H_{24}^* = \begin{bmatrix} H_{12}^* & H_{12}^* \\ H_{12}^* & -H_{12}^* \end{bmatrix}.$$

Let codes Q_{00}, Q_{01}, Q_{11} be formed as follows: Q_{00} consists of an all-zero word and an all-one word of length $n = 24$; the codewords of Q_{01} are rows of H_{24}^* and the codewords of Q_{01} are rows of H_{24}. Thus, we obtain $(24, 48|24)$-nonlinear PUM code with $d_{free} = 24, \alpha = 4$. It is clear that we can continue this process and obtain $(12i, 24i|12i)$-nonlinear PUM codes, $i \geq 1$, with $d_{free} = 12i, \alpha = 2i$.

2.2 Construction 2

Now we consider a nonlinear code with partial memory two. Let H_{12} be a Hadamard matrix of order 12 and let h_i be the i-th row of H_{12}. We form the codes Q_{00}, Q_{01}, Q_{02},
Q_{11}, Q_{22} as follows:

$$Q_{00} = \{(0,0,0,0,0,0,0,0,0,0,0,0), (1,1,1,1,1,1,1,1,1,1,1,1)\},$$

$$Q_{01} = \{h_0, h_1, h_2, h_3\},\ Q_{02} = \{h_0, h_4, h_5, h_6\},$$

$$Q_{11} = \{h_0, h_7, h_8, h_9\},\ Q_{01} = \{h_0, h_{10}, h_{11}, h_{10} + h_{11}\}.$$

From the results of [2] it follows that a free distance of a code with a quasi-generator matrix (2) can be estimated by the inequality

$$d_{free} \geq min\{d_m, d_s + d_e + 2d_o\},$$

where

$$d_m = d_{min} \begin{bmatrix} Q_{00} & & \\ Q_{01} & Q_{11} & \\ Q_{02} & & Q_{22} \\ & Q_{00} & \\ & Q_{01} & Q_{11} \\ & & Q_{00} \end{bmatrix},$$

$$d_s = d_{min} \begin{bmatrix} Q_{00} \\ Q_{01} \\ Q_{02} \end{bmatrix},\ d_e = d_{min} \begin{bmatrix} Q_{00} \\ Q_{11} \\ Q_{22} \end{bmatrix},\ d_o = d_{min} \begin{bmatrix} Q_{00} \\ Q_{01} \\ Q_{02} \\ Q_{11} \\ Q_{22} \end{bmatrix}.$$

In our case $d_m = 12, d_s = 4, d_e = 4, d_o = 2$, hence $d_{free} = 12$. It was found by computer calculation that $\alpha = 8/5$. Thus, we obtain a $(12, 32|16)$-nonlinear code with partial memory two with $d_{free} = 12, \alpha = 8/5$. Note that the best known linear code with partial memory two and $d_{free} = 12$ is the $(12, 16|8)$ code.

References

1. C. Thommesen and J. Justesen, "Bounds on Distance and Error Exponents of Unit Memory Codes," *IEEE Trans. Info. Theory*, vol.IT-29, pp.637–649, 1983.
2. U. Dettmar, U. K. Sorger, D. N. Gevorkian, and V. V. Ziablov, "An Extension of Partial Unit Memory Codes," *Proc., The Third International Colloquium on Coding Theory*, Dilijian, pp.14–17, 1990.
3. V. V. Ziablov, S. A. Shavgulidze, and M. G. Shanidze, "Bounds on Free Euclidean Distance for Various Systems of Modulation and Coding," *Problems of Control and Information Theory*, vol.17(6), pp.345–356, 1988.

On Periodic (Partial) Unit Memory Codes with Maximum Free Distance

V. Zyablov and V. Sidorenko

Institute for Problems of Information Transmission
Ermolovoy St. 19, Moscow 101447 GSP-4, Russia
e-mail: zyablov@ippi.msk.su sid@ippi.msk.su

Abstract. In this paper we describe two constructions of q-ary (partial) unit memory codes that achieve upper bounds on the free distance. These constructions are based on Reed-Solomon codes. We show that the increase α of the extended row distance of the proposed codes is larger than that for known ones.

1 Introduction

Let a minimal encoder of a q-ary $R = k/n$ convolutional code have m memory elements. The code is called a Partial Unit Memory (PUM) code if $m < k$, and it is called a Unit Memory (UM) code if $m = k$. We consider a slightly generalized definition of UM and PUM codes. The reason is that for given n and k both the upper bounds on free distance and the Viterbi decoding complexity depend on m only, not on the distribution of the memory over blocks.

There are two important characteristics of a convolutional code: the free distance d_{free} and the increase of the extended row distance α. The *extended row distance* d_l^r is defined [1] to be the minimum *Hamming* weight of all paths in the minimal code trellis that diverge from zero state only once and then return for the first time to the zero state after l branches. We call these paths *l-loops*. The *free distance* is defined as follows: $d_{\text{free}} = \min_{l=1,2,\ldots}\{d_l^r\}$. The α gives the *average linear increase* of d_l^r: $\alpha = \lim_{l\to\infty} d_l^r/l$.

The following upper bounds on the free distance and α are known [1, 2]. For a PUM code:

$$d_{\text{free}} \leq n - k + m + 1, \tag{1}$$

and for UM code:

$$d_{\text{free}} \leq 2n - k + 1. \tag{2}$$

The upper bound on α does not depend on the memory m of the code:

$$\alpha \leq n - k. \tag{3}$$

In sections 2 and 3 we will describe constructions of periodic PUM and UM convolutional codes that are based on Reed-Solomon (RS) codes. We show that these codes have maximal free distance (that they meet the upper bounds (1) and (2)) for the rates $R \leq 1/3$ and for $R > 1/2$. We also estimate the increase α for these codes. It does not achieve the upper bound (3) but in section 4 we will show that it is larger than that for known codes. The codes proposed in section 3 (G-construction) are UM codes in classic sense [1] also.

2 H-construction

Let us represent a parity check matrix H_{RS} of a $(2n, 2k)$ RS code (or shortened RS code) as $H_{RS} = \begin{pmatrix} A & B \\ C & D \end{pmatrix}$, where A, B, C, and D are $r \times n$ -submatrices, $r = n - k$, $q > 2n$. We define a convolutional code W by the following parity check matrix:

$$H = \begin{pmatrix} A & & & \\ C & D & & \\ & B & A & \\ & & C & D \\ & & & \dots \end{pmatrix}. \tag{4}$$

The code is a k/n convolutional code with memory $m = n-k$ by the construction.

Proposition 1. *The code defined by the parity check matrix* (4) *has the following parameters:*

$$\alpha \geq (n - k + 1)/2, \tag{5}$$

$$d_{\text{free}} \geq \begin{cases} 2(n-k)+1, & \text{if } R > 1/2, \\ 2(n-k)+2, & \text{otherwise.} \end{cases} \tag{6}$$

Corollary 2. *For $R = k/n > 1/2$ the code defined by the parity check matrix* (4) *is a PUM code with the free distance $d_{\text{free}} = 2(n-k)+1$ that meets the upper bound* (1).

Proof of Proposition 1. Let $\mathbf{w} = (\mathbf{w}_0, \mathbf{w}_1, \dots)$ denote a code word, where the block $\mathbf{w}_i$ is n-vector over $GF(q)$. Then

$$\mathbf{w}H^T = 0. \tag{7}$$

To estimate the extended row distance d_l^r, $l = 1, 2, \dots$, consider an l-loop

$$\mathbf{w}(j, l) = (\dots, 0, \mathbf{w}_j, \mathbf{w}_{j+1}, \dots, \mathbf{w}_{j+l-1}, 0, \dots). \tag{8}$$

Let $W(l)$ be the set of all possible l-loops, then

$$d_l^r = \min_{\mathbf{w}(j,l) \in W(l)} |\mathbf{w}(j, l)|.$$

Let us show that there are no zero blocks in the loop: $\mathbf{w}_i \neq 0$. The minimal trellis of the code W is isomorphic to the syndrome trellis [3] and can be constructed using the parity check matrix H (4) as follows. Let H be represented as $H = (H_1, H_2, \dots)$, where $H_1 = \begin{pmatrix} A \\ C \\ O \\ \dots \end{pmatrix}$, $H_2 = \begin{pmatrix} O \\ D \\ B \\ O \\ \dots \end{pmatrix}$, ..., and O is an

$(n-k) \times n$ zero matrix. The code word $\mathbf{w}$ passes through the state $\mathbf{s}_i(\mathbf{w})$ after the ith branch $\mathbf{w}_i$, where

$$\mathbf{s}_i(\mathbf{w}) = \mathbf{w}_1 H_1^T + \ldots + \mathbf{w}_i H_i^T. \tag{9}$$

It follows from (4), (7) and (9) that if $\mathbf{w}_i = 0$ then the branch $\mathbf{w}_i$ starts from the zero state and enters the zero state:

$$\mathbf{w}_i = 0 \quad \Rightarrow \quad \mathbf{s}_i(\mathbf{w}) = 0, \mathbf{s}_{i-1}(\mathbf{w}) = 0.$$

Hence, for the loop $\mathbf{w}(j,l)$ we have $\mathbf{w}_i \neq 0, i = j, j+1, \ldots, j+l-1$.

Let $d(E)$ denote the minimum Hamming distance of the block code having E as the parity check matrix. A code word $\mathbf{w}(j,1)$ satisfies either the relation $\mathbf{w}(j,1)\begin{pmatrix} A \\ C \end{pmatrix}^T = 0$ or $\mathbf{w}(j,1)\begin{pmatrix} D \\ B \end{pmatrix}^T = 0$. Hence

$$d_1^r = \min_{W(1)} |\mathbf{w}(i,1)| = d\begin{pmatrix} A \\ C \end{pmatrix} = d\begin{pmatrix} D \\ B \end{pmatrix} = \begin{cases} 2r+1, & \text{when } r < n/2 \\ \infty, & \text{otherwise.} \end{cases} \tag{10}$$

Similarly,

$$d_2^r = d\begin{pmatrix} A & \\ C & D \\ & B \end{pmatrix} = d\begin{pmatrix} D & \\ B & A \\ & C \end{pmatrix} \geq d(A) + d(B) = 2(r+1).$$

Here and later we use that $\mathbf{w}_i \neq 0$.

$$d_3^r = d\begin{pmatrix} A & & \\ C & D & \\ & B & A \\ & & C \end{pmatrix} \geq d(A) + d(BA) = 2(r+1).$$

Analogously,

$$d_l^r \geq \begin{cases} 2d(A) + d(BA)(l-2)/2 = (r+1)(l/2+1) & \text{if } l \text{ is even} \\ d(A) + d(BA)(l-1)/2 = (r+1)(l+1)/2 & \text{if } l \text{ is odd.} \end{cases}$$

Finally,

$$d_l^r \geq \left\lceil \frac{l+1}{2} \right\rceil (r+1), \quad l = 2, 3, \ldots, \tag{11}$$

and the estimations (5) and (6) follows from (10) and (11).

3 G-construction

Let us consider a sequence of $(2n, k)$ RS (or shortened RS) codes, $q > 2n$. Let the code C_i ($i = 0, 1, 2, \ldots$) be defined by the following list of roots of the generator polynomial:

$$\gamma^{is+1}, \gamma^{is+2}, \ldots, \gamma^{is+d-1},$$

where $d = 2n - k + 1$, s is a parameter, γ is a primitive element of the field $GF(q)$. Let G_i be a generator matrix of the code C_i and let G_i be represented as follows: $G_i = (G_{i0}, G_{i1})$, where G_{i0} and G_{i1} are $k \times n$ matrices. We define a convolutional code W by the following generator matrix:

$$G = \begin{pmatrix} G_{00} & G_{01} & & \\ & G_{10} & G_{11} & \\ & & G_{20} & G_{21} \\ & & & \cdots \end{pmatrix}. \tag{12}$$

The constructed code is a k/n UM convolutional code with memory $m = k$. Notice once more that the code is a UM one in the classic sense [1].

Proposition 3. *For $R = k/n \leq 1/3$ the code defined by* (11) *with $s = k$ has the following parameters*

$$d_{\text{free}} = 2n - k + 1; \qquad \alpha \geq n - 2k + 1. \tag{13}$$

The free distance of the code meets the upper bound (2). *The generator matrix of the code may be obtained by the periodic repetition of two matrices G_0, G_1.*

Proof. To estimate the extended row distance d_l^r, $l = 1, 2, \ldots$, consider an l-loop $\mathbf{w}(i, l)$ (see (8)). Without loss of generality let us consider the word $\mathbf{w}(0, l) = (\mathbf{w}_0, \ldots, \mathbf{w}_{l-1})$. The code word $\mathbf{w}(0, l)$ corresponds to the information sequence $\mathbf{u} = (\mathbf{u}_0, \mathbf{u}_1, \ldots, \mathbf{u}_{l-2})$, $\mathbf{w}(0, l) = \mathbf{u}G^\star$, where $G^\star$ consists of the first $l - 1$ rows of the matrix G, $\mathbf{u}_i \neq 0$ ($i = 0, \ldots, l - 2$). For $l = 1$ a loop $\mathbf{w}(0, 1)$ does not exist, hence $d_1^r = \infty$. For $l = 2$ we have $\mathbf{w}(0, 2) = \mathbf{u}_0 G_0$ ($\mathbf{u}_0 \neq 0$) is a nonzero codeword of the code C_0 and, hence,

$$d_2^r = |\mathbf{w}(0, 2)| \geq 2n - k + 1. \tag{14}$$

For simplicity let us assume that $q - 1 = 2n$. Thus, the codes C_i are nonshortened RS codes. The proof can be extended without difficulties for shortened RS codes too.

Let us show that for the loop $\mathbf{w}(0, l)$ it holds that $|\mathbf{w}_i| \geq n - k - s + 1 = n - 2k + 1$, $i = 1, \ldots, l - 2$. From (12) it follows that $\mathbf{w}_i = \mathbf{u}_{i-1} G_{i-1,1} + \mathbf{u}_i G_{i,0}$. Without loss of generality we may set $i = 1$ and consider $\mathbf{w}_1 = \mathbf{u}_0 G_{01} + \mathbf{u}_1 G_{10}$. Let us denote

$$\mathbf{c}_i = \mathbf{u}_i G_i = (\mathbf{u}_i G_{i,0}, \mathbf{u}_i G_{i,1}) = (\mathbf{c}_{i,0}, \mathbf{c}_{i,1}), i = 0, 1;$$

$$\mathbf{c}_1' = (\mathbf{c}_{11}, \mathbf{c}_{10}).$$

We have $\mathbf{w}_1 = \mathbf{c}_{01} + \mathbf{c}_{10}$, $\mathbf{c}_i \in C_i$, $\mathbf{c}'_1 \in C_1$, $\mathbf{c}'_1 \neq 0$ because $\mathbf{c}'_1$ is the cyclic shift of the nonzero codeword $\mathbf{c}_1$ and the code C_1 is cyclic.

To estimate $|\mathbf{w}_1|$ let us use the following relation [4]

$$|\mathbf{c}_0 + \mathbf{c}'_1| = |\mathbf{c}_{00} + \mathbf{c}_{11}| + |\mathbf{c}_{01} + \mathbf{c}_{10}| \leq n + |\mathbf{w}_1|. \tag{15}$$

Now we need to estimate $|\mathbf{c}_0 + \mathbf{c}'_1|$.

First, let us show that $\mathbf{c}_0 + \mathbf{c}'_1 \neq 0$. We will use the polynomial representation

$$c_0(x) = (x-\gamma)(x-\gamma^2)\dots(x-\gamma^{q-1-k})u_0(x),$$

$$c'_1(x) = (x-\gamma^{k+1})(x-\gamma^{k+2})\dots(x-\gamma^{q-1})u'_1(x),$$

where $\deg u_0(x) \leq k-1$, $\deg u'_1(x) \leq k-1$. If $c_0(x) + c'_1(x) = 0$ then $c_0(x)$ and $c'_1(x)$ must have the same roots. Thus, $c_0(x)$ must have $q-1$ roots $\gamma, \gamma^2, \dots, \gamma^{q-1}$. But $c_0(x)$ can not have $q-1$ roots since $\deg c_0(x) \leq q-2$.

Second, let us show that $|\mathbf{c}_0 + \mathbf{c}'_1| \geq 2n - 2k + 1$. Since $c_0(x)$ and $c'_1(x)$ have $q-1-2k$ common roots $\gamma^{k+1}, \dots, \gamma^{q-k-1}$, $c_0(x) + c'_1(x)$ is a nonzero codeword of RS code with minimal distance $q - 2k = 2n - 2k + 1$ and from (15) we have $|\mathbf{w}_1| \geq n - 2k + 1$.

Finally, consider $d_l^r = |\mathbf{w}_0, \mathbf{w}_1, \dots, \mathbf{w}_{l-1})|$, where $|\mathbf{w}_i| \geq n - 2k + 1$, $i = 2, \dots, l-2$. For the end blocks $\mathbf{w}_0$ and $\mathbf{w}_{l-1}$ we have the estimate $|\mathbf{w}_0| = |\mathbf{c}_{00}| = |(\mathbf{c}_{00}, \mathbf{c}_{01})| - |\mathbf{c}_{01}| \geq d - n = n - k + 1$. So,

$$d_l^r \geq 2(n-k+1) + (l-2)(n-2k+1), \quad l = 3, 4, \dots,$$

and using (14) we obtain for $k \leq n/3$ $d_{\text{free}} = \min_{l=1,2,\dots} d_l^r = 2n - k + 1$ and $\alpha = \lim_{l\to\infty} d_l^r / l = n - 2k + 1$.

In our proof we need to consider only two adjacent generator matrices G_{i-1} and G_i. Hence, to obtain generator matrix G of the convolutional code it is sufficient to repeat matrices G_0 and G_1 periodically. Q.E.D.

The G-construction provides a noncatastrophic codes for all rates, not only for $R \leq 1/3$. The estimation of d_{free} and α for rates $R > 1/3$ is in progress.

4 Discussion

Let us compare parameters of the proposed codes with those of time invariant codes [2, 5, 6], that also have optimal free distance.

In [5, 6] the PUM codes with optimal d_{free} that were considered that have $\alpha \geq n - k - m + 1$ $(k + m < n + 1)$. For $m = n - k$ it gives $\alpha \geq 1$ when the H-construction gives $\alpha \geq (n - k + 1)/2$. Notice, however, that the H-construction does not suggest codes with $m \neq n - k$, which the papers [5, 6] do. The papers [5, 6] also propose UM codes with the same $\alpha \geq n - 2k + 1$ as the G-construction (13), but $d_{\text{free}} = 2n - 2k + 2$ for these codes [5, 6] does not meet the upper bound (2).

In [6] the PUM code with optimal d_{free} was suggested with $\alpha \geq (n - k - m/2 + 1)/2$, but for $m = n - k$ the H-construction gives the greater value $\alpha \geq (n - k + 1)/2$.

PUM codes with optimal free distance were also suggested in [2] for $m \leq \min(k-1, n-k-1)$ but there are no estimates for α for these codes.

Thus, the proposed codes have the increase α larger than that of known codes with the same parameters n, k, m, and d_{free}.

References

1. C. Thommesen and J. Justesen, Bounds on distances and error exponents of unit memory codes, *IEEE Trans. Inform. Theory*, IT-29(5), 637-649, 1983
2. F. Pollara, R. J. McEliece, and K. Abdel-Ghaffar, Finite-states codes, *IEEE Trans. Inform. Theory*, IT-34(5), 1083-1089, 1988
3. V. V. Zyablov and V. R. Sidorenko, Bounds on complexity of trellis decoding of linear block codes, *Problems of Information Transmission*, 29(3), 3-9, Jul.-Sept., 1993 (transl. from Russian)
4. S. A. Popov, Synchronization of separate MDS codewords in presence of noise, *Problems of Information Transmission*, 21(3), 28-35, 1985 (transl. from Russian)
5. J. Justesen, Bounded distance decoding of unit memory codes, *IEEE Trans. Inform. Theory*, IT-39(5), 1616-1627, 1993
6. U. K. Sorger, A new construction of partial unit memory codes based on Reed-Solomon Codes, *Problems of Information Transmission*, submitted 1993.

Concatenated Codes with Convolutional Inner and Outer Codes

Uwe Dettmar[1] and Ulrich K. Sorger,[1] Victor. V. Zyablov[2]

[1] Institut für Netzwerk–und Signaltheorie
Technische Hochschule Darmstadt
Merckstr. 25, D–64283 Darmstadt, FR Germany
[2] Institute for Problem of Information Transmission
Russian Academy of Science
Ermolova 19, Moscow GSP-4, 101447
Russia

1 Introduction

The concept of (partial) unit memory ((P)UM) codes makes it possible to apply the theory of block codes to the construction and the decoding of convolutional codes. This also holds for the construction of concatenated codes. The basic tools for describing the properties of a concatenated system are the extended row and the extended column distance.

2 Definitions

In this paper we consider only binary codes. An encoder for an (n, k) unit memory code is defined by the following encoding rule:

$$\mathbf{y}_j = \mathbf{x}_j \mathbf{G}_0 + \mathbf{x}_{j-1} \mathbf{G}_1$$

with $\mathbf{x}_j$ a k-bit information vector, $\mathbf{y}_j$ an n-bit output vector and $\mathbf{G}_0$ and $\mathbf{G}_1$ $k \times n$ matrices. The rank of $\mathbf{G}_0$ is k and that of $\mathbf{G}_1$ is $k' \leq k$. If $k' < k$ then the codes are called $(n, k|k')$ partial unit memory codes. The encoding rule reads

$$\mathbf{y}_j = \mathbf{x}_j \begin{pmatrix} \mathbf{G}_{00} \\ \mathbf{G}_{01} \end{pmatrix} + \mathbf{x}_{j-1} \begin{pmatrix} \mathbf{G}_{10} \\ 0 \end{pmatrix},$$

where $\mathbf{G}_{00}$ and $\mathbf{G}_{10}$ are $k' \times n$ matrices and $\mathbf{G}_{01}$ is a $(k-k') \times n$ matrix. Let

$$\overline{\mathbf{x}} = \mathbf{x}_1, \mathbf{x}_2, .., \mathbf{x}_j, .. \tag{1}$$

be some sequence of input (information) blocks,

$$\overline{\mathbf{y}} = \mathbf{y}_1, \mathbf{y}_2, .., \mathbf{y}_j, .. \tag{2}$$

the corresponding sequence of output blocks, and let $w_H(\overline{\mathbf{y}})$ denote the Hamming weight of the sequences $\overline{\mathbf{y}}$.

Definition 1. Denote by I_j^r the set of information block sequences

$$\overline{\mathbf{x}} = \mathbf{x}_0, \mathbf{x}_1, \mathbf{x}_2, .., \mathbf{x}_i, .., \mathbf{x}_j, ...$$

that satisfy: $\mathbf{x}_0 = 0$; $\mathbf{x}_i \neq 0$ for $0 < i \leq j$ and $\mathbf{x}_{j+1} = 0$ elsewhere. Then the extended row distance $\hat{d}^r(j)$ is defined as follows:

$$\hat{d}^r(j) = \min_{\overline{\mathbf{x}} \in I_j^r} \{w_H\left(\overline{\mathbf{x}}_1 \mathbf{G}_{(P)UM}\right)\}, \tag{3}$$

with $\mathbf{G}_{(P)UM}$ denoting the generator matrix of the (P)UM code.

Definition 2. Denote by I_j^c the set of information block sequences

$$\overline{\mathbf{x}} = \mathbf{x}_0, \mathbf{x}_1, \mathbf{x}_2, .., \mathbf{x}_i, .., \mathbf{x}_j, ...$$

that satisfy: $\mathbf{x}_0 = 0$; $\mathbf{x}_i \neq 0$ for $0 < i \leq j$ and $\mathbf{x}_{j+1} = 0$ or $\mathbf{x}_{j+1} \neq 0$ elsewhere. Then the extended column distance $\hat{d}^{cs}(j)$ is defined as follows:

$$\hat{d}^{cs}(j) = \min_{\overline{\mathbf{x}} \in I_j^c} \{w_H\left(\overline{\mathbf{x}} \mathbf{G}_{(P)UM}\right)\}. \tag{4}$$

Definition 3. Denote by I_j^b the set of information block sequences

$$\overline{\mathbf{x}} = \mathbf{x}_0, \mathbf{x}_1, \mathbf{x}_2, .., \mathbf{x}_i, .., \mathbf{x}_j, ...$$

that satisfy: $\mathbf{x}_i \neq 0$ for $0 \leq i \leq j$ and $\mathbf{x}_{j+1} = 0$ elsewhere. Then the extended reverse column distance $\hat{d}^{ce}(j)$ is defined as follows:

$$\hat{d}^{ce}(j) = \min_{\overline{\mathbf{x}} \in I_j^b} \{w_H\left(\overline{\mathbf{x}} \mathbf{G}_{(P)UM}\right)\}. \tag{5}$$

The free distance d_∞ of a non-catastrophic (P)UM code is given by the minimum of the extended row distances:

$$d_\infty = \min_{j=0,1,...} \hat{d}^r(j).$$

In fact the minimum weight of an error burst of length j necessary to produce a decoding error of length j is given as $\left\lceil (\hat{d}^r(j)+1)/2 \right\rceil$, which illustrates the importance of this concept.

Example 1. For the binary (8, 4) PUM code found by Lauer [2] the generator matrices are

$$G_{00} = \begin{pmatrix} 1\,0\,0\,0\,1\,1\,0\,1 \\ 0\,1\,0\,0\,0\,1\,1\,1 \\ 0\,0\,1\,0\,1\,0\,1\,1 \end{pmatrix},$$
$$G_{01} = \begin{pmatrix} 1\,1\,1\,1\,1\,1\,1\,1 \end{pmatrix},$$
$$G_{10} = \begin{pmatrix} 1\,1\,0\,0\,1\,0\,0\,1 \\ 0\,1\,1\,0\,0\,1\,0\,1 \\ 1\,0\,1\,0\,0\,0\,1\,1 \end{pmatrix},$$

the extended row distance of this code is

$$\hat{d}^r(j) = \begin{cases} 8 & \text{for } j = 1 \\ 4 + 2 \cdot j & \text{for } j > 1 \end{cases},$$

and the extended column distance of this code is

$$\hat{d}^{cs}(j) = \hat{d}^{ce}(j) = 2 + 2 \cdot j \ \ for\ j \geq 1.$$

Example 2. For the (n_o, k_o) PUM code with memory v_o based on the (n_o, k_o) RS codes with minimum distance d_b constructed by Justesen [3] the parameters are

$$\begin{aligned} \hat{d}^r(j) &= 2 \cdot v_o + j \cdot (d_b - v_o), \\ \hat{d}^{cs}(j) &= \hat{d}^{ce}(j) = v_o + j \cdot (d_b - v_o) \end{aligned}.$$

If the $(15, 11)$ PUM code with memory $v_o = 2$ over $GF(2^4)$ is chosen, then the parameters are

$$\begin{aligned} \hat{d}^r(j) &= 4 + j \cdot 3, \\ \hat{d}^{cs}(j) &= \hat{d}^{ce}(j) = 2 + j \cdot 3 \end{aligned}$$

and $d_\infty = 7$.

3 Concatenated Codes with Outer Block and Inner Convolutional Codes

A typical encoder for a concatenated code is

$\Rightarrow$ [outer encoder] $\Rightarrow$ [buffer of outer code symbols] $\Rightarrow$ [inner encoder] $\Rightarrow$

The input of the encoder of the outer code is an array of information symbols. The input of the encoder of the inner code is determined by the output of the buffer, the size of the buffer, the order in which the symbols are written into the buffer and the order in which the symbols are read from buffer.

It is assumed that the alphabet of the outer code has the size 2^k with k the number of information symbols of the inner convolutional code as defined in the preceding section.

First a concatenated system without interleaving is considered. One has one outer code of distance d_o and the inner convolutional code with the given extended row distance $\hat{d}^r(j)$. That means that the buffer consists of one code word of the outer code:

$$\mathbf{x}_1, \mathbf{x}_2, .., \mathbf{x}_i, .., \mathbf{x}_{n_o},$$

with $\mathbf{x}_i$ the ith code symbol of the outer codeword. Each symbol $\mathbf{x}_i$ of the outer codeword is an input block of information symbols for the encoder of the inner convolutional code.

To estimate the minimum distance of such a scheme one states that

$$\hat{d}^r(j) + \hat{d}^r(i) \geq \hat{d}^r(j + i + 1) \tag{6}$$

holds for any good convolutional code. Furthermore, it is supposed that the mapping from the outer code symbols to the inner codewords is chosen in such a way that the resulting concatenated code is linear. The minimum weight of the concatenated codeword c is estimated in the following way:

$$\mathrm{wt}(\mathbf{c}) \geq \min_{\sum_{i=0}^{j} l_i = \mathrm{wt}(\mathbf{c_O})} \sum_{i=0}^{j} \hat{d}^r(l_i)$$

with $\mathbf{c_O}$ the outer codeword. The term becomes minimal for the minimal weight d_o and using (6)

$$d_c \geq \hat{d}^r(d_o)$$

with d_c the estimated concatenated distance.

3.1 Interleaving

If the code is interleaved the estimate of the concatenated distance improves. Let the code be encoded in the following way. First encode I linear outer codes of length n_o and distance d_o. Write these codewords in an $I \times n_o$ matrix (buffer), so that any row of this matrix is an outer codeword.

$$\begin{array}{c} \mathbf{x}_{11} \ldots \mathbf{x}_{1l} \ldots \mathbf{x}_{1n_o} \\ \ldots \ldots \ldots \ldots \ldots \\ \mathbf{x}_{I1} \ldots \mathbf{x}_{Il} \ldots \mathbf{x}_{In_o} \end{array}$$

with $\mathbf{x}_{il}$ the ith code symbol of the lth codeword. Use successively the outer symbols along a column of the matrix as information for the inner PUM Code. After one column is depleted take the next one and so on. The input sequence for the inner convolutional becomes

$$\mathbf{x}_{11} \ldots \mathbf{x}_{I1} \ldots \mathbf{x}_{1l} \ldots \mathbf{x}_{Il} \ldots \mathbf{x}_{1n_0} \ldots \mathbf{x}_{In_o}\,.$$

To estimate the concatenated distance, the minimum weight is again considered. If any outer codeword is the all zero word then the inner code is terminated at least d_o times. This means that the minimal weight is at least $d_o d_\infty$. If no code is terminated then the overall outer weight is at least Id_o and the weight becomes, with the same considerations as for no interleaving, at least $\hat{d}^r_{Id_o}$. Hence

$$d_c \geq \min\{d_o d_\infty, \hat{d}^r(Id_o)\}$$

with d_c the estimated concatenated distance. The estimated concatenated distance d_c will be increased with interleaving I until $d_o d_\infty \leq \hat{d}^r(I_0 d_o)$, with I_0 the optimal level of interleaving.

Example 3. Consider the binary concatenated code with the inner PUM code from example 1 and the outer RS code of length $n_o = 15$, $k_o = 11$ and $d_o = 5$ over $GF(2^4)$ and interleaving I. The estimated concatenated distance is

$$d_c \geq \hat{d}_c = \min\{5 \cdot 8,\ 4 + 2 \cdot I \cdot 5\},$$

or

I	1	2	3	4	5
$\hat{d}_c$	14	24	34	40	40
n_c	120	240	360	480	600
$\hat{d}_c/n_c$	0.117	0.1	0.094	0.083	0.067

4 Concatenated Codes with Inner and Outer Convolutional Codes

The approach used for outer block codes also applies for the case of outer convolutional codes, but the resulting code will be a convolutional code. As for outer block codes it is assumed that

- the alphabet of the outer code has the size 2^{k_i} with k_i the number of information symbols of the inner convolutional code as defined in the preceding section;
- the inner code is a linear binary (partial) unit memory (n_i, k_i) code with free distance $d_{i\infty}$, extended row distance $\hat{d}_i^r(l)$ and extended column distance $\hat{d}_i^c(l)$;
- the outer code is a linear q-ary (partial) unit memory (n_o, k_o) code with free distance $d_{o\infty}$, extended row distance $\hat{d}_o^r(l)$ and extended column distance $\hat{d}_o^c(l)$;.

The main aim of the investigation is the estimation of the free distance $d_{c\infty}$ of the concatenated convolutional code and how this estimation depends on the interleaving.

First no interleaving is assumed. That means that in the buffer with size $1 \times n_o$ the sequence of information blocks are

$$\mathbf{x}_1^1 \ldots \mathbf{x}_{n_o}^1, \mathbf{x}_1^2 \ldots \mathbf{x}_{n_o}^2, \ldots, \mathbf{x}_1^l \ldots \mathbf{x}_{n_o}^l, \ldots,$$

where $\mathbf{x}_j^l$ is the jth code symbol of the lth block of the outer code, which is sent to the encoder of the inner code. This sequence of information blocks contains at least $d_{o\infty}$ nonzero blocks. The worst case (6) is that all of these blocks occur in one burst. The free concatenated distance $d_{c\infty}$ is therefore estimated by

$$d_{c\infty} \geq \hat{d}_i^r(d_{o\infty}).$$

Note that the block length n_c of the convolutional concatenated code is equal to

$$n_c = n_i \cdot n_o \, .$$

4.1 Interleaving

To include interleaving is a more difficult task. This is the case because both the outer and the inner code are convolutional codes and need to be interleaved to increase the free distance.

Firsdt consider interleaving of only the inner code. The buffer size is then equal to $I \times n_o$. The buffer is given by

$$\begin{matrix} \mathbf{x}_1^1 \dots \mathbf{x}_{n_o}^1 \\ \dots \dots \dots \\ \mathbf{x}_1^I \dots \mathbf{x}_{n_o}^I \end{matrix}, \quad \begin{matrix} \mathbf{x}_1^{I+1} \dots \mathbf{x}_{n_o}^{I+1} \\ \dots \dots \dots \\ \mathbf{x}_1^{I+I} \dots \mathbf{x}_{n_o}^{I+I} \end{matrix}, \dots$$

where $\mathbf{x}_j^l$ is the jth code symbol of the lth block of the outer code. The sequence of information blocks of the inner code becomes

$$\mathbf{x}_1^1\, \mathbf{x}_1^2 \dots \mathbf{x}_1^I\, \mathbf{x}_2^1 \dots \mathbf{x}_2^I \dots$$

One worst case (6) is that in which all of these blocks are in one burst, which occupies $\hat{d}_o^{cs}(1)$ adjacent columns.

$$\begin{matrix} \# \dots \# \# \\ \dots \dots \dots \\ \# \dots \# \# \\ \# \dots \# \, 0 \end{matrix},$$

where # - denote some nonzero information block for the inner encoder (nonzero symbol of the outer code). The free distance of the concatenated code can be estimated by

$$d_{c\infty} \geq \min\{\hat{d}_i^r(\hat{d}_o^{cs}(1) \cdot I - 1),\; d_{i\infty} \cdot \hat{d}_o^{cs}(1)\} \text{ for } I > 1.$$

In this case the block length n_c of the convolutional concatenated code is equal to

$$n_c = n_i \cdot n_o \cdot I\,.$$

Because the length of the concatenated codes increases linearly with I, the ratio between n_c and the estimated concatenated distance $\hat{d}_c$ will be the same with increasing interleaving I until $\hat{d}_i^r(\hat{d}_o^{cs}(1) \cdot I_0 - 1) = d_{i\infty} \cdot \hat{d}_o^{cs}(1)$ and I_0 the optimal level of interleaving for this case.

Example 4. Consider the binary convolutional concatenated code with the inner PUM code from example 1 and the outer PUM code from example 2 and interleaving I of the inner code. The estimated concatenated distance is

$$d_c \geq \hat{d}_c = \min\{4 + 2 \cdot (I \cdot 5 - 1),\; 8 \cdot 5\},$$

or

I	1	2	3	4	5
$\hat{d}_c$	18	22	32	40	40
n_c	120	240	360	480	600
$\hat{d}_c/n_c$	0.15	0.092	0.089	0.083	0.067

Now consider interleaving of both the inner and the outer code. This means that the buffer size is equal to $I \times Jn_o$. The buffer is given by the matrix

$$\begin{array}{cccccccccc} \mathbf{x}_1^1 & .. & \mathbf{x}_{n_o}^1 & .... & \mathbf{x}_1^J & .. & \mathbf{x}_{n_o}^J & \mathbf{x}_1^{IJ+1} & .... & \mathbf{x}_1^{IJ+J} .. \mathbf{x}_{n_o}^{IJ+J} \\ .. & .. & .. & ..... & .. & .. & .., & ... & .... & .. \quad .., ... \\ \mathbf{x}_1^{(I-1)J+1} & .. & \mathbf{x}_{n_o}^{(I-1)J+1} & .... & \mathbf{x}_1^{IJ} & .. & \mathbf{x}_{n_o}^{IJ} & \mathbf{x}_1^{(2I-1)J+1} & .... & \mathbf{x}_1^{2IJ} .. \mathbf{x}_{n_o}^{2IJ} \end{array}$$

where $\mathbf{x}_j^l$ is the jth code symbol of the lth block of the outer code. In this case the sequence of information blocks becomes

$$\mathbf{x}_1^1\, \mathbf{x}_1^{J+1} \ldots \mathbf{x}_1^{(I-1)J+1}\, \mathbf{x}_2^1 \ldots \mathbf{x}_2^{(I-1)J+1} \ldots$$

The worst case (6) is that in which all of these blocks are in one burst, which occupies $\widehat{d}_o^{cs}(J)$ adjacent columns. The worst case becomes

$$\begin{array}{c} \# .. \# \# \\ ... . \\ \# .. \# \# \\ \# .. \# \; 0 \end{array}$$

Then the free distance of the concatenated code can be estimated by

$$d_{c\infty} \geq \min\{\widehat{d}_i^r(\widehat{d}_o^{cs}(J) \cdot I - 1),\ d_{i\infty} \cdot d_{o\infty}\} \text{ for } I > 1.$$

In this case the block length n_c of the convolutional concatenated code is equal to

$$n_c = n_i \cdot n_o \cdot I\, J.$$

Because the length of the concatenated codes increases linearly with the product IJ, the ratio between n_c and the estimated concatenated distance d_c is minimized by increasing the interleaving IJ until $\widehat{d}_i^r(\widehat{d}_o^{cs}(J) \cdot I - 1) = d_{i\infty} \cdot d_{o\infty}$. Note that this estimation of the free distance of the concatenated code is very rough. For $J > 2$ the worst case that all $\widehat{d}_o^{cs}(J)$ are adjacent columns, does not exist and the real free distance of the concatenated code will be better.

Example 5. Consider the binary convolutional concatenated code with the inner PUM code from example 1 and the outer PUM code from example 2 and interleaving I of the inner code and interleaving of the outer code $J = 2$. The estimated concatenated distance is

$$d_c \geq \widehat{d}_c = \min\{4 + 2 \cdot (I \cdot 7 - 1),\ 8 \cdot 7\},$$

or

I	1	2	3	4	5
$\widehat{d}_c$	18	30	44	56	56
n_c	120	480	720	960	1200 ..
$\widehat{d}_c/n_c$	0.15	0.063	0.061	0.058	0.047

Note that for interleaving $I = 4$ and $J = 2$ the estimated concatenated distance is equal to

$$d_{c\infty} = d_{i\infty} \cdot d_{o\infty}\ .$$

5 Summary

In this paper a short description of concatenated codes with inner convolutional codes is given. The distance parameters of such codes can be estimated using the extended row and column distances of the inner and outer codes.

References

1. J. Justesen, C. Thommesen, and V. Zyablov, "Concatenated codes with inner convolutional codes," *IEEE Trans. Inform. Theory*, vol. IT-34, pp. 1217–1225, Sept. 1988.
2. G.S.Lauer, "Some optimal partial-unit-memory codes," *IEEE Trans. Inform. Theory*, vol. IT-25, pp. 240-243, Mar. 1979.
3. J.Justesen, "Bounded distance decoding of unit memory codes," *IEEE Trans. Inform. Theory, vol. IT-39, pp. 1616-1627, Sept. 1993.*

Reduced-State Decoding for Trellis Coded Modulation on Nonlinear Intersymbol Interference Channels

Felix A. Taubin

St.-Petersburg State Academy of Airspace Instrumentation,
Bolshaia Morskaia str., 67, St.-Petersburg, 190000, Russia,
e-mail:liap@sovam.com

1 Introduction

The combination of trellis coding and modulation for nonlinear intersymbol interference (ISI) channels provides an essential improvement in performance over conventional transmission schemes. The optimum decoding strategy for trellis codes operating over nonlinear intersymbol interference channels is maximum likelihood (ML) sequence estimation. Possible noiseless output channel sequences can be represented by an ML trellis which is constructed by combining the encoder states with the states, created by nonlinear ISI. Then this trellis is searched with the Viterbi algorithm. The main problem with the optimum decoder is its complexity, for the number of nodes of the ML trellis is typically too large.

To overcome this complexity various processors have been suggested to shorten channel memory by using channel equalization (or ISI-state truncation techniques) and to pass the resulting output to the Viterbi decoder. (See for example McLane [1], Wesolowski [2], Eyuboğly [3], Chevillat and Eleftheriou [4]). Another approach to reducing the Viterbi decoder's complexity has been described by Biglieri [5] and Marsan *et al.* [6]. For a fixed channel memory a subset of state variables are chosen for reduced-state or mismatched decoding, and a receiver operates with the Viterbi algorithm whose metric is based on a partial (or null) knowledge of the channel. This approach leads to a family of Viterbi decoders which offer a tradeoff between decoding complexity and performance.

In this paper a more general reduced-state decoding scheme is considered. Reduced-state (RS) trellises are constructed by merging the states of the ML trellis. Each path in the RS trellis is regarded as an approximation of original path in the ML trellis, and an assignment for the branch metric is based on the minimum square error criterion. A performance bound for the reduced-state decoder is derived. This bound differs from the one for the ML decoder in that a new parameter (called as the quasi-distance) plays the role of the Euclidean distance. A lower bound on the minimum quasi-distance used to evaluate the asymptotic performance is given.

2 System Description

A transmission system partitions the binary information sequence into blocks of length k and sends a k-block $a_i = (a_{i1},...,a_{ik})$ per i-th signaling interval. The sequence $\mathbf{a} = (a_0, a_1,...)$ is encoded by a trellis encoder with the rate k/n bit/coded symbol and the constraint length ν. The output from this encoder is the sequence $\mathbf{c} = (c_0, c_1,...)$ of n-blocks $c_i = (c_{i1}, ..., c_{in})$, $c \in \mathcal{C}$. The trellis encoder produces a block of coded symbols c_i each time a block of information bits a_i enters the encoder: $c_i = c(a_i,...,a_{i-\nu})$, where $c(\cdot,...,\cdot)$ is a coding rule. The n-blocks $c_0, c_1,...$ are subsequently mapped into N-dimensional channel symbols $b_0, b_1,...$ taken from a 2^n–ary signal constellation $B \subset \mathbf{C}^N$ in accordance with a memoryless mapping rule $\varphi(\cdot)$. The rate of the mapper is therefore n coded symbols/channel symbol, and the overall rate is k information bit/channel symbol.

The channel symbol sequence $\mathbf{b} = (b_0, b_1,...)$ is the input to a modulator. The transmitted signal is

$$s(t) = \text{Re}\,[u(t)\exp(j2\pi f_0 t)], \quad t \geq 0,$$

where $u(\cdot)$ is the equivalent low-pass signal waveform (the complex envelope of $s(\cdot)$) and f_0 denotes the carrier frequency. The equivalent low-pass signal $u(\cdot)$ has the form

$$u(t) = \sqrt{2E_s}\sum_{j\geq 0} u(b_j, t - jT),$$

where $\{u(b,\cdot) \mid b \in B\}$ is a set of basic waveforms, $1/T$ is the symbol rate and E_s is the transmitted energy per channel symbol. Assume that, for each $b \in B$, the waveform $u(b,\cdot)$ has duration T. Then

$$\int_0^T \overline{|u(b,t)|^2}\,dt = 1,$$

where overbar indicates an average over the ensemble of channel symbols. The energy per information bit $E_b = E_s\,/k$.

The signal $s(t)$ is transmitted over a nonlinear band-limited channel having an equivalent low-pass impulse response $h(\cdot)$ and memoryless nonlinearity $F(\cdot)$. The equivalent low-pass received signal has the form

$$r(t) = F\left(\int_0^t h(\tau)u(t\text{-}\tau)d\tau\right) + n(t)$$

$$= F\left(\sqrt{2E_s}\sum_{j\geq 0} g(b_j, t - jT)\right) + n(t),$$

where channel response

$$g(b_j,\cdot) = u(b_j,\cdot) * h(\cdot),$$

$*$ denotes the convolution operation, and $n(\cdot)$ represents the additive white Gaussian noise with two-sided power spectral density N_0.

Assume that the convolution $g(\cdot,\cdot)$ is characterized by a finite duration of length $L+1$ symbol intervals; i.e. $L+1$ is the smallest integer such that $\forall b \in B$ $g(b,t) = 0$ for $t \geq (L+1)T$. Consequently, the channel memory is equal to L.

3 Maximum Likelihood Decoding

The response $g(b,\cdot)$ may be regarded as the sequence of L T-finite chips $g_l(b,\cdot)$, $0 \leq l \leq L$, i.e.

$$g(b,t) = \sum_{l=0}^{L} g_l(b, t - lT), \quad 0 \leq t \leq (L+1)T,$$

where

$$g_l(b,t) = \begin{cases} g(b, lT + t), & t \in [0,T], \\ 0, & \text{otherwise.} \end{cases}$$

Then the received energy per channel symbol is

$$E = 1/2 \int_0^T \left| F(\sqrt{2E_s} \sum_{l=0}^{L} g_l(\varphi(c(a_{j-l}, \ldots, a_{j-l-\nu})), t) \right|^2 dt.$$

Denoting

$$v(a_j, \ldots, a_{j-\nu-L}, \cdot) = F\left(\sqrt{2E_s} \sum_{l=0}^{L} g_l(\varphi(c(a_{j-l}, \ldots, a_{j-l-\nu})), \cdot)\right) / \sqrt{2E},$$

the equivalent low-pass output of the channel is

$$r(t) = \sqrt{2E} \sum_{j \geq 0} v(a_j, \ldots, a_{j-\nu-L}, t - jT) + n(t).$$

It is obvious that the cascade of encoder and channel can be seen as a finite state machine with states s_j given by

$$s_j = (a_{j-1}, \ldots, a_{j-\nu-L});$$

each state s_j takes on values from an alphabet $\mathcal{S}$ of $2^{k(\nu+L)}$ states S_m, $1 \leq m \leq 2^{k(\nu+L)}$. The output signal of this machine, corresponding to a transition from state s_j to state s_{j+1}, is $z(s_j, s_{j+1}, \cdot) = v(a_j, \ldots, a_{j-\nu-L}, \cdot)$.

Let H be the linear span of the set $\{z(S_m, S_l, \cdot) \mid S_m, S_l \in \mathcal{S}\}$, let h be the dimension of H, and let $\{f_i(\cdot) \mid 1 \leq i \leq h\}$ be an orthonormal basis of the linear span H. The outputs of a bank of h matched filters, one matched to each basis function, in the jth timing instant are

$$r_{jl} = \int_{jT}^{(j+1)T} r(t) f_l^*(t\text{-}jT)dt = \sqrt{2E} z_l(s_j, s_{j+1}) + n_{jl}, \quad 1 \leq l \leq h,$$

where $f_l^*(t)$ denotes the complex conjugate of $f_l(t)$,

$$z_l(\mathrm{s}_j, \mathrm{s}_{j+1}) = \int_0^T z(\mathrm{s}_j, \mathrm{s}_{j+1}, t) f_l^*(t) dt,$$

$$n_{jl} = \int_{jT}^{(j+1)T} n(t) f_l^*(t\text{-}jT)dt.$$

Denoting $\mathbf{r}_j = (r_{j1},\ldots,r_{jh})$, $\mathbf{z}_j = (z_1(\mathrm{s}_j,\mathrm{s}_{j+1}),\ldots,z_h(\mathrm{s}_j,\mathrm{s}_{j+1}))$, $\mathbf{n}_j = (n_{j1},\ldots,n_{jh})$, the output of the bank is $\mathbf{r}_j = \sqrt{2E}\mathbf{z}_j + \mathbf{n}_j$. The sequence $\mathbf{r} = (\mathbf{r}_0, \mathbf{r}_1, \ldots)$ represents a set of sufficient statistics for determining the information sequence $\mathbf{a}$. Notice that the set $\mathcal{Z}$ of all sequences $\mathbf{z} = (\mathbf{z}_0, \mathbf{z}_1, \ldots)$ may be regarded as an "output" code unlike the "input" code $\mathcal{C}$.

A ML decoder determines among all possible information sequences the sequence $\hat{\mathbf{a}}$ that maximizes the metric

$$\Gamma(\mathbf{a}) = \sum_{j\geq 0} \sum_{l=1}^{h} \left(2\mathrm{Re}\left[r_{jl}^* z_l(s_j, s_{j+1}) \right] - \sqrt{2E} |z_l(s_j, s_{j+1})|^2 \right);$$

it operates on the combined code $\mathcal{C}$ and the ISI trellis (ML trellis), which has $2^{k(\nu+L)}$ nodes. The bit error probability can be upper-bounded by using union-bound arguments:

$$P_b \leq \sum_{\mathbf{a}} \Pr[\mathbf{a}] \sum_{\mathbf{a}' \in \mathcal{E}(\mathbf{a})} d_H(\mathbf{a}, \mathbf{a}') \Pr[\mathbf{a} \mapsto \mathbf{a}']. \tag{1}$$

In (1) $\Pr[\mathbf{a} \mapsto \mathbf{a}']$ is the pairwise probability, $d_H(\mathbf{a}, \mathbf{a}')$ is the Hamming distance between the information sequences $\mathbf{a}$ and $\mathbf{a}'$, $\mathcal{E}(\mathbf{a})$ is the set of all possible detours (error events) with respect to sequence $\mathbf{a}$ starting at time 0, and $\Pr[\mathbf{a}]$ is the probability of $\mathbf{a}$. The pairwise probability is

$$\begin{aligned} \Pr[\mathbf{a} \mapsto \mathbf{a}'] &= Q(d(\mathbf{a}, \mathbf{a}')\sqrt{E/2N_0}) \\ &\leq \exp(-d^2(\mathbf{a}, \mathbf{a}')E/4N_0), \end{aligned} \tag{2}$$

where

$$d(\mathbf{a}, \mathbf{a}') = \left(\sum_{j\geq 0} \sum_{l=1}^{h} |z_l(s_j, s_{j+1}) - z_l(s_j', s_{j+1}')|^2 \right)^{1/2}$$

and $Q(x)$ is Q-function , the area under the unit normal curve from x to infinity.

To get an upper bound in the standard form, it is necessary to introduce the product trellis and generalized transfer function (see Biglieri [5]). Let $\mathcal{O}$ be the set of product states (s, s') such that $s = s'$, and let $A(d, v, l, o)$ be the number of detours $\mathbf{e} = ((a_0, a_0^{'}, ..., (a_{l-1}, a_{l-1}^{'}))$ in the product trellis starting at the node $o \in \mathcal{O}$ and having length l, Hamming weight $d_H(\mathbf{e}) = d_H(\mathbf{a}, \mathbf{a}') = v$ and Euclidean weight $d^2(\mathbf{e}) = d^2(\mathbf{a}, \mathbf{a}') = d^2$. Also, let $\mathcal{D}$ be the set of all possible $d(\mathbf{e})$. The generalized transfer function is defined as

$$T(D, I) = 2^{-k(\nu+L)} \sum_{d \in \mathcal{D}} \sum_{v} \sum_{l \geq \nu+L} 2^{-kl} \sum_{o \in \mathcal{O}} A(d, v, l, o) D^{d^2} I^v. \qquad (3)$$

Substituting (2) in (1) and using definition (3), the upper bound on the bit error probability can be expressed as

$$P_b \leq k^{-1} \frac{dT(D, I)}{dI} \bigg|_{\exp(-E/4N_0),\quad I=1}.$$

At high signal-to-noise ratios the bit error probability can be estimated by the term involving the minimum distance $d_{\min} = \min\{d(\mathbf{e}) | d(\mathbf{e}) \in \mathcal{D}\}$:

$$P_b \simeq N(d_{min}) \exp\left(-(E/4N_0) d_{min}^2\right),$$

where

$$N(d_{\min}) = 2^{-k(\nu+L)} \sum_{v} v \sum_{l \geq \nu+L} 2^{-kl} \sum_{o \in \mathcal{O}} A(d_{\min}, v, l, o)$$

is the error coefficient of the "output" code $\mathcal{Z}$.

4 Reduced-State Decoding

Let us introduce a finite state machine with states σ_j given by

$$\sigma_j = (\mathrm{a}_{j-1}, \ldots, \mathrm{a}_{j-\mu}), \qquad \mu < \nu + L;$$

each state σ_j takes on values from an alphabet $\mathcal{R}$ of $2^{k\mu}$ states Σ_m, $1 \leq m \leq 2^{k\mu}$. Obviously, the state σ_j represents the union of $2^{k\rho}$ states s_j, $\rho = \nu + L - \mu$, having the same first μ k–tuples. The output T–finite signal of this machine, corresponding to a transition from state σ_j to state σ_{j+1}, is denoted by $y(\sigma_j, \sigma_{j+1}, \cdot)$. The overall output signal

$$\sum_{j \geq 0} y(\sigma_j, \sigma_{j+1}, t - jT), \quad t \geq 0,$$

can be considered as some approximation of the signal

$$\sum_{j \geq 0} z(s_j, s_{j+1}, t - jT), \quad t \geq 0.$$

It is more realistic to suppose that impairment of the error performance due to merging of the states of the ML trellis is significantly determined by the mean square error of approximation. Given the sequence **a**, the mean square error of approximation

$$\sup_i (1/i) \sum_{j=0}^{i-1} \int_0^T |z(s_j, s_{j+1}, t) - y(\sigma_j, \sigma_{j+1}, t)|^2 dt$$

can be upper—bounded as

$$\sup_i (1/i) \sum_{j=0}^{i-1} \max_{(s_j, s_{j+1}) \in (\sigma_j, \sigma_{j+1})} \int_0^T |z(s_j, s_{j+1}, t) - y(\sigma_j, \sigma_{j+1}, t)|^2 dt.$$

Therefore for each pair of successive states (Σ_p, Σ_q) define $y(\Sigma_p, \Sigma_q, \cdot)$ as the function that minimizes the maximum square error

$$\max_{(S_m, S_l) \in (\Sigma_p, \Sigma_q)} \int_0^T |z(S_m, S_l, t) - y(\Sigma_p, \Sigma_q, t)|^2 dt \tag{4}$$

corresponding to this pair, and denote

$$\Delta(\Sigma_p, \Sigma_q) = \min_{y(\Sigma_p, \Sigma_q, \cdot)} \max_{(S_m, S_l) \in (\Sigma_p, \Sigma_q)} \int_0^T |z(S_m, S_l, t) - y(\Sigma_p, \Sigma_q, t)|^2 dt.$$

Let W be the linear span of the set $\{y(\Sigma_m, \Sigma_l, \cdot) \mid \Sigma_m, \Sigma_l \in \mathcal{R}\}$, let w be the dimension of W, and let $\{\psi_i(\cdot) \mid 1 \leq i \leq w\}$ be an orthonormal basis of the linear span W. The outputs of a bank of w matched filters in the jth timing instant are

$$r_{jl} = \int_{jT}^{(j+1)t} r(t) \psi_l^*(t - jT) dt = \sqrt{2E} y_l(\sigma_j, \sigma_{j+1}) + n_{jl}, \quad 1 \leq l \leq h,$$

where

$$y_l(\sigma_j, \sigma_{j+1}) = \int_0^T y(\sigma_j, \sigma_{j+1}, t) \psi_l^*(t) dt, \qquad n_{jl} = \int_{jT}^{j+1} n(t) \psi_l^*(t\text{-}jT) dt.$$

Denoting $\mathbf{r}_j = (r_{j1}, \ldots, r_{jw})$, $\mathbf{y}_j = (y_1(\sigma_j, \sigma_{j+1}), \ldots, y_w(\sigma_j, \sigma_{j+1}))$, $\mathbf{n}_j = (n_{j1}, \ldots, n_{jw})$, the output of the bank is $r_j = \sqrt{2E} y_j + n_j$. The set $\mathcal{Y}$ of all sequences $\mathbf{y} = (\mathbf{y}_0, \mathbf{y}_1, \ldots)$ may be regarded as a "simplified" image of the code $\mathcal{Z}$. It is obvious that the distinction between $\mathcal{Y}$ and $\mathcal{Z}$ increases as the number of states of the RS trellis is decreased.

A reduced-state decoder determines the information sequence $\hat{\mathbf{a}}$ that maximizes the metric

$$\overline{\Gamma}(\mathbf{a}) = \sum_{j\geq 0}\sum_{l=1}^{w}\left(2\mathrm{Re}\left[r_{jl}^{*}y_l(\sigma_j,\sigma_{j+1})\right] - \sqrt{2E}|y_l(\sigma_j,\sigma_{j+1})|^2\right);$$

it operates on the RS trellis, which has $2^{k\mu}$ nodes. The bit error can be upper-bounded in the form (1). But now the pairwise probability

$$\Pr[\mathbf{a}\mapsto\mathbf{a}'] = Q(\overline{d}(\mathbf{a},\mathbf{a}')\sqrt{E/2N_0}),$$

where

$$\overline{d}(\mathbf{a},\mathbf{a}') = \left(\sum_{j\geq 0}\sum_{l=1}^{w}\mathrm{Re}[(2z_l(s_j,s_{j+1}) - y_l(\sigma_j,\sigma_{j+1}) - y_l(\sigma_j',\sigma_{j+1}'))^*\right.$$

$$\left.\times(y_l(\sigma_j,\sigma_{j+1}) - y_l(\sigma_j',\sigma_{j+1}'))]\right)\left(\sum_{j\geq 0}\sum_{l=1}^{w}|y_l(\sigma_j,\sigma_{j+1}) - y_l(\sigma_j',\sigma_{j+1}')|^2\right)^{-1/2}.$$

If the function $\overline{d}(\cdot,\cdot)$ is nonnegative for all pairs $(\mathbf{a},\mathbf{a}')$, it can be interpreted as a quasi-distance. Defining the minimum quasi-distance

$$\overline{d}_{\min} = \min_{\mathbf{a},\mathbf{a}'\neq\mathbf{a}}\overline{d}(\mathbf{a},\mathbf{a}'),$$

the bit error probability can be asymptotically estimated as

$$P_b \simeq N(\overline{d}_{\min}\exp(-(E/4N_0)\overline{d}_{\min}^2),$$

where $N((\overline{d})_{\min}$ is the error coefficient of the code $\mathcal{Y}$. The asymptotic power loss of the reduced-state decoder in reference to the ML decoder is given by

$$G = 10\log_{10}\left(d_{min}^2/\overline{d}_{\min}^2\right).$$

Clearly, the minimum quasi-distance is a non-decreasing function of the parameter μ, and $\overline{d}_{min}$ coincides with d_{min} when μ is equal to $\nu+L$.

5 A Lower Bound on the Minimum Quasi-Distance

Let us denote

$$q(t) = \sum_{j\geq 0}(z(s_j,s_{j+1},t-jT) - z(s_j',s_{j+1}',t-jT)),$$

$$\varepsilon_{1j}(t) = z(s_j,s_{j+1},t-jT) - y(\sigma_j,\sigma_{j+1},t-jT), \quad \varepsilon_1(t) = \sum_{j\geq 0}\varepsilon_{1j}(t),$$

$$\varepsilon_{2j}(t) = z(s_j^{'}, s_{j+1}^{'}, t - jT) - y(\sigma_j^{'}, \sigma_{j+1}^{'}, t - jT), \qquad \varepsilon_2(t) = \sum_{j\geq 0} \varepsilon_{2j}(t).$$

Then $\bar{d}(\mathbf{a}, \mathbf{a}')$ can be represented in the form

$$\bar{d}(\mathbf{a}, \mathbf{a}') = \frac{< (q(\cdot)+\varepsilon_1(\cdot)+\varepsilon_2(\cdot)),(q(\cdot)-\varepsilon_1(\cdot)-\varepsilon_2(\cdot)) >}{\|q(\cdot) - \varepsilon_1(\cdot) - \varepsilon_2(\cdot)\|}, \tag{5}$$

where $< \cdot,\cdot >$ and $\|\cdot\|$ are the scalar product and norm on $L^2(0, \infty)$ respectively.

The following simple lemma will be needed.

Lemma 1. For any $q(\cdot),\varepsilon(\cdot) \in L^2(0,\infty)$

$$\frac{< (q(\cdot)+\varepsilon(\cdot)),(q(\cdot)-\varepsilon(\cdot)) >}{\|q(\cdot) - \varepsilon(\cdot)\|} \geq \|q(\cdot)\| - \|\varepsilon(\cdot)\|.$$

Applying this lemma to (5) and taking into account that

$$\|\varepsilon_{ij}(\cdot)\|^2 \leq \max_{(\Sigma_p,\Sigma_q)} \Delta(\Sigma_p, \Sigma_q), \qquad i = 1, 2,$$

one obtains

$$\begin{aligned}
\bar{d}(\mathbf{a}, \mathbf{a}') &\geq d(\mathbf{a}, \mathbf{a}') - \|\varepsilon_1(\cdot)\| - \|\varepsilon_2(\cdot)\| \\
&\geq d(\mathbf{a}, \mathbf{a}') - \left(\sum_{j\geq 0} \|\varepsilon_{1j}(\cdot)\|^2\right)^{1/2} - \left(\sum_{j\geq 0} \|\varepsilon_{2j}(\cdot)\|^2\right)^{1/2} \\
&\geq d(\mathbf{a}, \mathbf{a}') - 2\left(|J| \max_{(\Sigma_p,\Sigma_q)} \Delta(\Sigma_p, \Sigma_q)\right)^{1/2},
\end{aligned}$$

where $J = \{j \mid a_j \neq a^{'}\}$, and $\mid J \mid$ denotes cardinality of the set J.

An upper bound on $\Delta(\Sigma_p,\Sigma_q)$ is given by

Lemma 2. Let $y(\Sigma_p,\Sigma_q,\cdot)$ be defined as the function that minimizes the maximum square error (4). Then $\Delta(\Sigma_p,\Sigma_q)$ can be approximated as

$$\Delta(\Sigma_p, \Sigma_q) \leq \max_{(S_m,S_l),(S_k,S_n)\in(\Sigma_p,\Sigma_q)} \beta \int_0^T |z(S_m, S_l, t) - z(S_k, S_n, t)|^2 dt$$

with $\beta = 2^{k\rho-1}/(2^{k\rho} + 1)$.

In this way $\bar{d}(\mathbf{a}, \mathbf{a}')$ becomes lower-bounded as

$$\bar{d}(\mathbf{a}, \mathbf{a}') \geq d(\mathbf{a}, \mathbf{a}') - 2\sqrt{|J|\Delta} \tag{6}$$

with

$$\Delta = \max_{(\Sigma_p,\Sigma_q)} \max_{(S_m,S_l),(S_k,S_n)\in(\Sigma_p,\Sigma_q)} \beta \int_0^T |z(S_m, S_l, t) - z(S_k, S_n, t)|^2 dt.$$

In the special case that the channel is linear ($F(\cdot)$ is the identity mapping),

$$z(s_j, s_{j+1}, \cdot) = \sum_{l=0}^{L} g_l(\varphi(c(a_{j-l}, \ldots, a_{j-l-\nu})), \cdot)$$

and hence

$$\Delta = \max_{a_1,\ldots,a_\nu} \max_{\substack{a_{\nu+1},\ldots,a_{\nu+\rho} \\ a'_{\nu+1},\ldots,a'_{\nu+\rho}}} \left(\int_0^T |\sum_{l=1}^{\rho} (g_{l+\mu-\nu}(\varphi(c(a_l,\ldots,a_{l+\nu})),t) \right.$$

$$\left. -g_{l+\mu-\nu}(\varphi(c(a'_l,\ldots,a'_{l+\nu})),t))|^2 dt \quad |a'_i = a_i, \quad 1 \le i \le \nu \right).$$

Taking into account that for every detour $\overline{\mathbf{e}}$ in the product RS trellis there is a corresponding detour $\mathbf{e}$ in the product ML trellis and using the lower bound (6), one obtains the net result:

$$\overline{d}_{\min} \ge \min_{e}(d(\mathbf{e}) - 2\sqrt{l(\mathbf{e})\Delta},$$

where $l(\mathbf{e})$ denotes the length of $\mathbf{e}$.

References

1. P.J. McLane, "A residual intersymbol interference error bound for truncated-state Viterbi detectors", IEEE Trans on Inf. Theory, vol. IT-26, pp.548-553, September, 1980.
2. K. Wesolowski, "Efficient digital receiver structure for trellis-coded signals transmitted through channels with intersymbol interference," Electron. Lett., pp.1265-1267, Nov. 1987.
3. M.V. Eyuboğlu, "Detection of coded modulation signals on linear severely distorted channels using detection-feedback noise prediction with interleaving," IEEE Trans. Commun., vol. COM-36, pp.401-409, Apr. 1988.
4. P.R. Chevillat and E. Eleftheriou, "Decoding of trellis-encoded signals in presence of intersymbol interference and noise," IEEE Trans. Commun., vol. COM-37, pp.669-676, July 1989.
5. E. Biglieri, "High level modulation and coding for nonlinear satellite channels," IEEE Trans. Commun., vol. COM-32, pp.616-626, May 1984.
6. M.A. Marsan, G. Albertengo and S. Benedetto, "Combined coding and modulation: performance over real channels," in Digital Communications, E. Biglieri and G. Prati, Eds. Elsevier Science Publishers B.V. (North-Holland), pp.71-82, 1986.

Tables of Coverings for Decoding by S-Sets

I.L.Asnis, S.V.Fedorenko, E.A.Krouk, E.T.Mironchikov

St.-Petersburg State Academy of Airspace Instrumentation,
Bolshaia Morskaia str., 67, St.-Petersburg, 190000, Russia,
e-mail: liap@sovam.com

Abstract. A linear block decoding method in which decoding is reduced to the repeated decoding of shortened codes is proposed. The algorithm for constructing coverings from codewords, which are necessary for the proposed technique, is provided. A tables of such coverings for a number of binary BCH and quadratic-residue (QR) codes is presented.

1 Notation

A linear block decoding method called S-sets decoding was suggested in [1]. Let us adopt the following notation. Let $\mathcal{G}$ be a linear (n, k, d)-code of length n, dimension k and minimum Hamming distance d. Code $\mathcal{G}$ has a generator matrix G and a parity-check matrix H. Let us enumerate the positions of codewords numerically from 1 to n. Let us call the set of numbers

$$J = \{j_1, \ldots, j_s\}, \qquad 1 \le j_1 < \ldots < j_s \le n,$$

an S-set. Denote the subvector of vector $f = (f_1, \ldots, f_n)$ with numbers from the set J as $f(J) = (f_{j_1}, \ldots, f_{j_s})$, and the submatrix of matrix $M = [m_1; \ldots; m_n]$ as $M(J) = [m_{j_1}; \ldots; m_{j_s}]$, where m_i is the i-th column of matrix M. Then the matrix $G(J)$ determines $(S, rank\ G(J))$-code $\mathcal{G}(J)$ with minimum distance d_J. Let I_f be the set of zero positions of the vector f and let $|V|$ be the power of the set V. Let $[x]$ be an integer not larger than x and let $W(f)$ be the Hamming weight of the vector f. A vector $b = a + e$ is a received vector, where $a \in \mathcal{G}$ and e is an error vector. For decoding it is enough to have the family of S-sets $J : \{J_i\}, (i = \overline{1, Q})$, for which it is possible to find for any $e \in E$ at least one subvector $e(J_i)$ which is decoded correctly by the code $\mathcal{G}(J_i)$, where

$$E = \{e : W(e) \le [(d-1)/2]\}.$$

If the vector $e(J)$ is decoded correctly in the code $\mathcal{G}(J)$ then the set J is the decoding set for e.

2 Decoding by S-sets

The basic idea behind the proposed decoding method (decoding by S-sets, or S-decoding) is as follows. For any codeword a, the subvector $a(J)$ is a codeword of the code $\mathcal{G}(J)$. If J is the vector's e decoding set, then by decoding $b(J)$ in the code $\mathcal{G}(J)$ one may determine $a(J)$, and then by $a(J)$, one may also find vector a. So it is possible now to interpret the decoding process as the process of decoding set search.

The described approach generalizes to information sets decoding - the search concludes with the determination that the information set is free from error.

Let $\hat{a}(J)$ be the result of the decoding of the vector $b(J)$ in the code $\mathcal{G}(J)$. Let us take into account the vector b with a list $L_J(b)$ of the code $\mathcal{G}$ codewords coinciding with $\hat{a}(J)$ in any positions from J. If there are not any words $a \in \mathcal{G}$ such that $a(J) = \hat{a}(J)$ then this list may be empty. If J is the information set then $|L_J(b)| = 1$.

Theorem 1. *To be a decoding S-sets J it is necessary and sufficient that*

$$\min_{a \in L_J(b)} d(a,b) \leq t,$$

where $t = [(d-1)/2]$ *and* $d(a,b)$ *is the Hamming distance between the vector* a *and* b.

Let us now formulate the decoding by S-sets algorithm.

1. For each S-set $J_i \in \{J_i\}$ do the following steps.

1.1. Decode the vector $b(J_i)$ in code $\mathcal{G}(J_i)$ for S-set J_i. The vector $\hat{a}(J_i)$ is determined by the decoding operation.

1.2. The list of codewords $L_J(b) = \{\hat{a}\} \in \mathcal{G}$ that coincide with $\hat{a}(J_i)$ in any positions from J_i is constructed.

1.3. The vectors $\hat{a}$ for which $d(\hat{a},b) \leq [(d-1)/2]$, form $\{\hat{a}\}$, where the vector $\hat{a}$ is a decoded variant of the received word.

2. If a vector $\hat{a}$ is not found for any of the S-set from $\{J_i\}$, an uncorrectable error is assumed to have occurred.

The decoding complexity is determined by the family $\{J_i\}$ power. As we construct the decoders by S-sets, the construction of a minimal power family $\{J_i\}$ is the central issue.

For a family $\{J_i\}$ to be sufficient for S-sets decoding, it is necessary and sufficient that for any vector $e \in E$ the inequality

$$W(e(J_i)) \leq \left[\frac{d_{J_i}-1}{2}\right]$$

is true for some $J_i \in \{J_i\}$.

3 Coverings Consisting of Codewords

Special attention is paid to the construction technique based on shortened codes for the family $\{J_i\}$, as described by Helgert and Stinaff [2].

If $a \in \mathcal{G}$ and $W(a) < 2d$ then residual code $\mathcal{G}(I_a)$ has the parameters $(n - W(a), k-1, d_a)$, where $d_a \geq d - [W(a)/2]$.

We consider the connection of the family $\{J_i\}$ construction problem to Turan's problem [3]. Turan's covering $T(n,t,\tau)$ is a set of τ-subsets of set $\{1,2,\ldots,n\}$ such that any t-subset of set $\{1,2,\ldots,n\}$ includes at least one τ-subset. Covering $M(n,t,\tau)$ is a set of such t-subsets of set $\{1,2,\ldots,n\}$ such that any τ-subset of set $\{1,2,\ldots,n\}$ is a subset of some t-subset.

$T(n,t,\tau)$ usually consists of all τ-subsets V_i, $i = \overline{1,l}$, such that

$$\bigcup_{i=1}^{l} V_i = \{1,2,\ldots,n\}.$$

By use of S-sets it is possible to organize the decoding algorithm for the code $\mathcal{G}$ through repeated decoding of the residual codes. For the decoding algorithm we should have a set of codewords $\{a_1,\ldots,a_Q\} \in \mathcal{G}$ such that for any error vector $e \in E$ it is possible to find a word $a_i \in \mathcal{G}, (1 \leq i \leq Q)$, such that

$$W(e(I_{a_i})) \leq \left[\frac{d_{a_i}-1}{2}\right].$$

We propose the algorithm for constructing coverings from codewords for decoding by S-sets [4]. This algorithm consist of two steps.

1. First, Turan's covering with parameters

$$T(n,[(d-1)/2],[(d-1)/2]-[(d_a-1)/2])$$

is constructed. This covering consists of all $([(d-1)/2]-[(d_a-1)/2])$-subsets of set V_i, $i = \overline{1,l}$.

2. The second step is to find codewords forming coverings

$$M(|V_i|, W(a_i), [(d-1)/2]-[(d_a-1)/2])$$

on each set V_i, $i = \overline{1,l}$, which were determined during Turan's covering construction. These words can be found as words of code $\mathcal{C}$ with parity-check matrix $H(V_i)$ and parameters

$$(|V_i|, |V_i| - rank\ H(V_i), d_{V_i}),$$

where $d_{V_i} \geq d$.

Parameters for coverings consisting of codewords for several QR and BCH codes are cited in Table 1. The weight of the codewords is denoted as W. Residual codes have parameters (n_a, k_a, d_a). The number of residual codes (used for decoding) is denoted by Q.

Let us try to explain the contents of Table 1. For example, the fourth string shows that for decoding the (42,21,10)-code it is sufficient to decode 16 residual codes: 15 (32,20,5)-codes and 1 (30,20,4)-code. Coverings for the codes 1, 2, 7, 8, 9, 12, 13, 15 are optimum for decoding by S-sets. Coverings for codes 3, 11, 14 are optimum for the proposed algorithm. The covering for code 5 is an optimum one for this algorithm if Turan's hypothesis about the numbers

$$|T(2L,5,3)| = 2 * \binom{L}{3}$$

is true.

Table 1. Parameters of coverings consisting of codewords for QR and BCH codes.

N	(n,k,d)	W	(n_a,k_a,d_a)	Q
1	18,9,6	6	12,8,3	3
2	24,12,8	8	16,11,4	6
3	32,16,8	8	24,15,4	12
4	42,21,10	10	32,20,5	15
		12	30,20,4	1
5	48,24,12	12	36,23,6	28
6	48,24,12	16	32,23,4	87
7	15, 5,7	7	8, 4,4	1
		8	7,4,3	1
8	15, 7,5	5	10, 6,3	3
9	31, 6,15	15	16, 5,8	1
		16	15, 5,7	1
10	31,11,11	11	20,10,6	3
		12	19,10,5	6
		16	15,10,3	3
11	31,16,7	7	24,15,4	3
		8	23,15,3	9
12	31,21,5	5	26,20,3	7
13	63, 7,31	31	32, 6,16	1
		32	31, 6,15	1
14	63,24,15	15	48,23,8	15
		16	47,23,7	45
15	63,51,5	5	58,50,3	13

Though the proposed decoding algorithm doesn't give the asymptotic decrease of estimated linear codes decoding complexity as compared with the one obtained in [5], it is represented as a prospective decoding method for moderate length codes. The use of shortened codes for decoding was suggested by I.I. Dumer [6,7].

4 Example

Now we can describe the decoding algorithm for the Golay code. Let us denote the (24,12,8) Golay code as $\mathcal{G}_{24}$. One can show that for any a ($a \in \mathcal{G}_{24}$, $W(a) = 8$) there exists a residual code $\mathcal{G}(I_a)$ with parameters (16,11,4). Let the set of codewords $\{a_1, \ldots, a_Q\}$, ($a \in \mathcal{G}_{24}$, $W(a) = 8$), cover any two errors, i.e. it is a covering such that for any error $e = (e_1, \ldots, e_{24})$, $W(e) \leq 3$, it is possible to find $e(I_i)$, such that $W(e(I_i)) \leq 1$. Let us describe the decoding algorithm using the Helgert-Stinaff residual codes. First, for any a_i from the covering we construct the set $\mathrm{I}_{a_i} \equiv I_i$, the code $\mathcal{G}(I_i)$, the information set γ_i ($|\gamma_i| = k - 1$), the parity-check matrix H_i and the table of syndromes $S_i = b(I_i)H_i^T$ and corresponding coset leaders for the residual code $\mathcal{G}(I_i)$.

Then we construct the matrix $\hat{G}_i$, which is needed for decoding the i-th residual code. For this we write the generator matrix G of the code $\mathcal{G}_{24}$ in the form

$$G = \left\| \frac{a_i}{G_i} \right\|,$$

where a_i is the codeword and G_i is the $(k-1) \times n$ matrix.

Let us denote the code with generator matrix G_i as $\mathcal{G}_i$ and construct in this $(n, k-1)$-code $\mathcal{G}_i$ an information set γ_i such that $\gamma_i \subset I_i$. This can be done because $\mathcal{G}_i(I_i) \equiv \mathcal{G}(I_i)$ and $rank\ G(I_i) = k - 1$. Then $\hat{G}_i = [G_i(\gamma_i)]^{-1}G_i$. Let us enumerate the columns of the parity-check matrix $H_i : H_i = [h_{i,1}; \ldots; h_{i,16}]$, where $h_{i,j}$ is the j-th column of H_i.

Then we set for any syndrome $\mathrm{S}_i = h_{i,j}$ the corresponding coset leader $e(I_i) = (0 \ldots 0e_j0 \ldots 0)$, $e_j = 1$.

The algorithm consist of 6 steps.

For $i = \overline{1, Q}$ do

1. Calculate $S(I_i) = b(I_i)H_i^T$.
2. If $S(I_i) = (0 \ldots 0)$ then $e(I_i) = (0 \ldots 0)$; if $S(I_i) = h_{i,j}$ then $e(I_i) = (0 \ldots 0e_j0 \ldots 0)$, $e_j = 1$.
3. Calculate the codeword of the i-th residual code $\mathcal{G}(I_i)$: $q(I_i) = e(I_i) + b(I_i)$.
4. Calculate the codeword $p_i = (q(I_i)(\gamma_i))\hat{G}_i$ of the code $\mathcal{G}_i$. Operator $q(\cdot)(\cdot)$ denotes the calculation subvector of the subvector.
5. If $d(p_i, b) \leq 3$ then $\hat{a} = p_i$ is the decoded codeword;
   if $d(p_i + a_i, b) \leq 3$ then $\hat{a} = p_i + a_i$.
6. If the conditions in step 5 are false for all i then an uncorrectable error is assumed to have occurred.

References

1. Krouk E.A., Mironchikov E.T., Fedorenko S.V. "Decoding by S-sets." *In: Fifth Joint Soviet-Swedish International Workshop on Information Theory "Convolutional Codes; Multi-User Communication"*, Moscow, 1990, pp.113–115.

2. Helgert H.J., Stinaff R.D. "Minimum-distance bounds for binary linear codes", *IEEE Trans. Info. Theory*, vol. IT-19, pp.344–356, 1973.
3. Erdős P., Spencer J. Probabilistic methods in combinatorics. *Akad. Kiado*, Budapest, 1974.
4. Asnis I.L., Fedorenko S.V. "Tables of coverings for decoding by *S*-sets." *In: The Workshop on Information Protection*, Moscow, 1993, p.22.
5. Krouk E.A. "A bound on the decoding complexity of linear block codes." *Problemy Peredachi Informatsii*, 1989, vol.25, No.3, pp.103–107. (in Russian)
6. Dumer I.I. Private communication, 1986.
7. Dumer I.I. "On minimum distance decoding of linear codes." *In: Fifth Joint Soviet-Swedish International Workshop on Information Theory "Convolutional Codes; Multi-User Communication"*, Moscow, 1990, pp.50–52.

Periodicity of One-Dimensional Tilings

V. Sidorenko

Institute for Problems of Information Transmission
Ermolovoy St. 19, Moscow 101447 GSP-4, Russia
e-mail: sid@ippi.msk.su

Abstract. The tiling problem is closely connected with a number of problems of information theory. We show that the Wang-Moore conjecture is valid for the one dimensional case. Namely, any set of templates which permits a tiling of the set $\mathbb{Z}$ of integers also permits a periodic tiling. We also show that any tiling of $\mathbb{Z}$ by one template is necessarily periodic. The obtained results are also valid for tiling of the set $\mathbb{R}$ of real numbers.

1 Introduction

Suppose we are given a figure (template) T on the plane. We are interested in tiling (covering) of the plane by disjoint copies of the template T obtained by translations only (without rotations). It was shown in [1],[2],[3] that the tiling problem is closely connected with the following problems of information theory.

1. Let code words be $n \times n$-matrices. We assume that one error can corrupt a code word's elements inside any translation of the template T. If the template T tiles the plane then the code correcting $d/2$ errors can be constructed [1] by special interleaving of codes with distance d (Hamming metric). Some optimal codes can be obtained using this construction.

2. Another problem is connected with array processing. The effective utilization of the array processor depends on being able to arrange the data elements of arrays in parallel memory modules so that memory conflicts are avoided when the data are fetched. Let the template T describe the matrix subparts most often demanded in numerical computations. A conflict-free storage allocation strategy can be proposed [2] if the template T tiles the plane.

3. Finally, consider the problem of memory testing. For a defective memory chip, the state of any memory cell may depend on the states of some neighboring cells that belong to the region of fixed configuration. Let the configuration be defined by the template T. If the template tiles the set E then an optimal test may be constructed [3].

A practical implementation of these problems is strongly simplified when the tiling is periodic. This is why the periodicity of tilings is an important question.

Let us give some definitions. Consider the set $E = \mathbb{Z}^n$ of n-tuples of integers. A *template* T is a finite set $T = \{t_1, t_2, \ldots, t_w\}$, $t_i \in \mathbb{Z}^n$, $|T| = w$. A *translation* $T + u$ of the template T by vector $u \in \mathbb{Z}^n$ is $T + u = \{t_1 + u, t_2 + u, \ldots, t_w + u\}$.

A finite set of templates $\mathbf{T} = \{T_1, T_2, \ldots, T_s\}$ *tiles* the set $\mathbb{Z}^n$ if there exist the sets $U_i \subset \mathbb{Z}^n$, $i = 1, \ldots, s$, such that

$$E = \mathbb{Z}^n = \bigcup_{i=1}^{s} \bigcup_{u \in U_i} T_i + u, \tag{1}$$

and the covering sets in (1) are disjoint:

$$T_i + u_i \cap T_j + u_j = \emptyset, \quad i \neq j, u_i \in U_i, u_j \in U_j;$$

$$T_i + u_i \cap T_i + u_j = \emptyset, \quad u_i \neq u_j, u_i, u_j \in U_i.$$

The set $\mathbf{U} = \{U_1, U_2, \ldots, U_s\}$ is called the **T**-*tiling of the set* E.

The tiling $\mathbf{U}$ is called *periodic* if the vectors $v_1, \ldots, v_n$; $v_i \in \mathbb{Z}^n$ exist such that

$$U_i + v_j = U_i; \quad i = 1, \ldots, s; j = 1, \ldots, n,$$

and $\operatorname{rank}(v_1, \ldots, v_n) = n$.

Example 1. Let $n = 1$, $E = \mathbb{Z}$. Given the set $\mathbf{T}$ of two templates $\mathbf{T} = \{T_1, T_2\}$, where $T_1 = \{0\}, T_2 = \{0, 1\}$:

$T_1 =$ [0] $\qquad T_2 =$ [0 1].

Consider the following **T**-tiling of $E = \mathbb{Z}$: $\mathbf{U} = (U_1, U_2)$, where $U_1 = \{3i\}_{i \in \mathbb{Z}}$, $U_2 = \{3i + 1\}_{i \in \mathbb{Z}}$. The tiling may be drawn as follows:

−8 −7	−6	−5 −4	−3	−2 −1	0	1 2	3	4 5	6	7 8

This tiling is periodic because there exists the vector $v = (3)$ such that $U_i + v = U_i$, $i = 1, 2$. The tiling picture will not change after shifting by 3 cells.

Let us consider another **T**-tiling of $\mathbb{Z}$ (symmetrical with respect to 0):

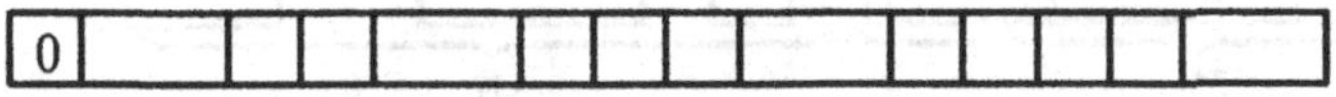

It is clear that this tiling is not periodic.

Moore [4] and Wang [5] made the following conjecture.

Conjecture 1 Wang, Moore. *Any finite set* $\mathbf{T}$ *of templates which permits a tiling of the plane* $\mathbb{Z}^2$ *also permits a periodic tiling.*

Berger [6] proved that the conjecture is false in general. He showed that there must be some set of templates (more than 20,000) which can tile the plane, but can not tile the plane periodically. The number of templates needed for Berger's nonperiodicity theorem was then reduced to 52 by Robinson [7].

On the other hand it was shown in [8] that the conjecture is valid for tiling the plane by one *polyomino* [1]. Moreover, the polyomino permits a regular tiling, that is the set U is a lattice [8]. Any tiling of the plane by one polyomino is half-periodic [9].

In this paper it will be shown that Conjecture 1 is valid for the one dimensional set $\mathbb{Z}$. Moreover, any tiling of $\mathbb{Z}$ by one template is necessarily periodic. These results were also generalized for tiling of $\mathbb{R}$ (set of real numbers). In this case a template is defined to be a finite set of intervals.

The tiling periodicity problem was also studied in physics [10], [11]. Perhaps, some similar results can be derived from [11], but the proposed paper still seems to be interesting because of its very simple proofs.

2 Main Results

In this section we will consider tilings of the one dimensional set $E = \mathbb{Z}$. Without loss of generality we will assume that the first element $t_{i,1}$ of a template $T = \{t_{i,1}, t_{i,2}, \ldots, t_{i,w_i}\}$, $t_{i,1} < t_{i,2} < \ldots < t_{i,w_i}$ is equal to zero: $t_{i,1} = 0$. Let $l(T_i) = t_{i,w_i} - 1$ be the length of the template T_i, and let $l = l(\mathbf{T})$ be the maximum length of the templates in $\mathbf{T}$: $l(\mathbf{T}) = \max_{T_i \in \mathbf{T}} l(T_i)$.

Theorem 2. *Any finite set* $\mathbf{T}$ *of templates which permits a tiling of the set* $\mathbb{Z}$ *also permits a periodic tiling.*

Proof. Let $\mathbf{U}$ be the $\mathbf{T}$-tiling of $\mathbb{Z}$. Given the integer μ, consider the interval $M = (-\infty, m_0] \subset \mathbb{Z}$. Deleting from the tiling $\mathbf{U}$, translations of templates that do not intersect with M_0, we obtain a set $\mathbf{U}_0^{(1)}$ such that the set

$$N_1 = \bigcup_{i=1}^{s} \bigcup_{u \in U_i^{(1)}} T_i + u$$

covers the interval M, $N_1 \supseteq M$ as shown below.

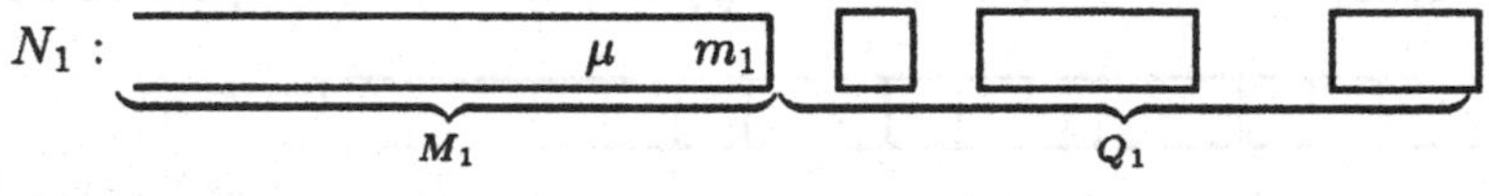

Let m_1 be the maximum integer such that N_1 covers the interval $(-\infty, m_1] = M_1$, $N_1 \supseteq M_1$, $m_1 + 1 \notin N_1$. The rest of the set $N_1 \setminus M_1 = Q_1$ is called a tail. The tail Q_1 consists of the integers from the interval $[m_1+2, m_1+l-1]$. Since each of $l-2$ integers from this interval may belong or do not belong to the tail Q_1, the number of different tails Q_1 is at most 2^{l-2}. Let us draw the set $N_1 = M_1 \cup Q_1$ symbolically as follows.

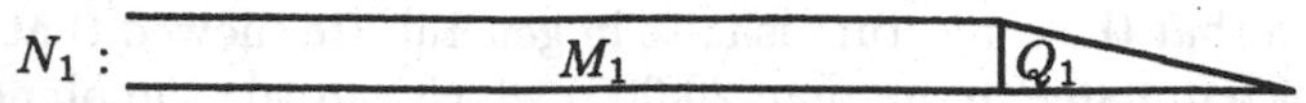

[1] A polyomino [4] is a template of which the cells form a rook-wise connected set with no "holes".

Deleting from the set N_1 the translation of a template T_i+u that contains m_1 we obtain the set $N_2 = N_1 \setminus (T_i + u)$, $\mathbf{U}^{(2)} = \mathbf{U}^{(1)} \setminus u$. Let m_2 be the maximum integer such that $(-\infty, m_2] \subseteq N_2$, $m_2 + 1 \notin N_2$. We denote $M_2 = (-\infty, m_2]$, $Q_2 = N_2 \setminus M_2$. Notice, that $m_1 - m_2 \leq l$.

Let us continue the deleting procedure. We will obtain the infinite sequence of tails $Q_1, Q_2, \ldots$. There are at least two identical (up to the shift) tails Q_i and Q_j in the sequence $Q_1, \ldots, Q_{2^{l-2}+1}$; $Q_j = Q_i + r$; $r \leq l2^{l-2}$. Hence $N_j = N_i + r$.

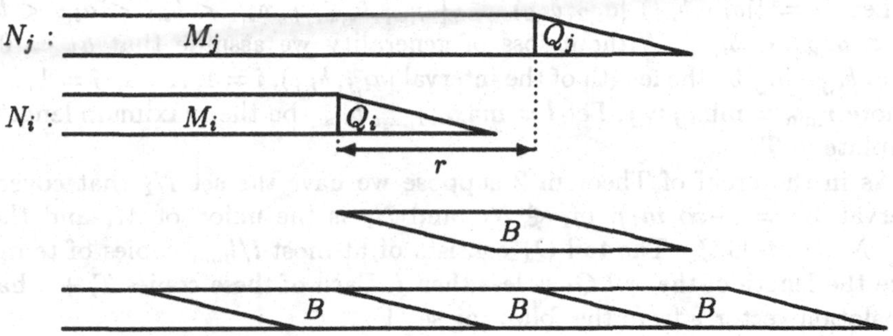

Consider now the set $B = N_j \setminus N_i = (N_i + r) \setminus N_i$. Let us show that a tiling of $\mathbb{Z}$ can be obtained by periodic repetition of the set B. Indeed, set B and $(B + r)$ fit to each other, i.e.

$$(B + r) \cap B = (N_i + 2r) \setminus (N_i + r) \cap (N_i + r) \setminus N_i = \emptyset,$$

$$(B + r) \cup B = (N_i + 2r) \setminus (N_i + r) \cup (N_i + r) \setminus N_i = (N_i + 2r) \setminus N_i,$$

since $N_i \subset N_i + r \subset N_i + 2r \subset \ldots$ is the sequence of nested sets. Thus, the set $\mathbb{Z}$ can be tiled by the periodic repetition of the set B. Since, there exists the **T**-tiling $\mathbf{U}^{(j)} \setminus \mathbf{U}^{(i)}$ of the set B itself, we obtain the periodic **T**-tiling of $\mathbb{Z}$. The period of the tiling is at most $l2^{l-2}$.

Theorem 3. *Any tiling of $\mathbb{Z}$ by a single template T is necessarily periodic.*

Proof. We are given the tiling U of the set $\mathbb{Z}$ by the template T. As in the proof of Theorem 1 consider the set $U^{(1)} \subset U$ that tiles the interval $M_1 = (-\infty, m_1]$, $N_1 \supseteq M_1$, $m_1 + 1 \notin N_1$, $N_1 = M_1 + Q_1$. Let us continue the tiling $U^{(1)}$ to obtain the tiling of $\mathbb{Z}$. This continuation is unique and, hence, it coincides with the tiling U. Indeed, to continue N_1 we have to cover the number $m_1 + 1$ by $T + u$ without overlapping with M_1. The only possibility is to place the leftmost cell $t_1 = 0$ of the template T at the place $m_1 + 1$, that is $u = m_1 + 1$. The continuation is periodic because the number of tails is finite. The period is at most $l2^{l-2}$ (see proof of Theorem 1). Hence, the original tiling U is also periodic.

3 Generalizations

In this section we will show that Theorems 2 and 3 may be generalized for tilings of the set $E = \mathbb{R}$.

Let $E = \mathbb{R}$ be the set of real numbers. In this case a template is a finite set of half-open intervals $T = \{[a_1, b_1), [a_2, b_2), \dots, [a_w, b_w)\}$ and $\mathbf{T} = \{T_1, \dots, T_s\}$.

Theorem 4. *Any finite set* $\mathbf{T}$ *of templates which permits a tiling of the set* $\mathbb{R}$ *also permits a periodic tiling.*

Proof. The proof is similar to one of Theorem 2. We need to show only that the number of different tails is finite when we disjoint our tiling.

Let $T_i = \{[a_{i,1}, b_{i,1}), [a_{i,2}, b_{i,2}), \dots, [a_{i,w_i}, b_{i,w_i})\}$, $a_{i,1} < b_{i,1} < a_{i,2} < b_{i,2} < \dots < a_{i,w_i} < b_{i,w_i}$. Without loss of generality we assume that $a_1 = 0$. Let $r_{i,j} = b_{i,j} - a_{i,j}$ be the length of the interval $[a_{i,j}, b_{i,j})$, $i = 1, \dots, s$; $j = 1, \dots, w_i$. Denote $r_{\min} = \min_{i,j} r_{i,j}$. Let $l = \max_{i=1,\dots,s} b_{i,w_i}$ be the maximum length of a template in $\mathbf{T}$.

As in the proof of Theorem 2 suppose we have the set N_1 that covers the interval $M_1 = (-\infty, m_1)$, $m_1 \notin N_1$, and N_1 is the union of M_1 and the tail $Q_1 : N_1 = M_1 \cup Q_1$. The tail Q_1 consists of at most $l/l_{\min}$ copies of templates since the length of the tail Q_1 is less then l. Each of these copies $T_i + v$ has the translation vector v from the following set V:

$$V = \{v : v = m_1 - \sum r_{i,j}, \quad v > m_1 - l\}, \tag{2}$$

since the set $(-\infty, m_1)$ is tiled. The number of addends in (2) is restricted by $l/r_{\min}$ because $v > m_1 - l$. Hence,

$$|V| \leq w_\sigma^{l/r_{\min}},$$

where $w_\sigma = \sum_{i=1}^{s} w_i$ is the total number of intervals in the templates $T_1, \dots, T_s$. So, we have at most $l/r_{\min}$ templates in the tail Q_1 and each is translated by a vector from V. Hence, the number of different tails is at most

$$|V|^{l/r_{\min}} \leq w_\sigma^{(l/r_{\min})^2},$$

and the tiling can be made periodic with period at most $l w_\sigma^{(l/r_{\min})^2}$. Q.E.D.

Theorem 5. *Any tiling of* $\mathbb{R}$ *by a single template* T *is necessarily periodic.*

The proof of the theorem is the same as the one for Theorem 3.

In conclusion notice that Theorem 2 can also be generalized for tilings of the space $\mathbb{Z} \times M_1 \times \dots \times M_n$ where $M_i = \{1, \dots, m_i\}$.

References

1. V. Sidorenko, Tilings of the plane and codes for translational combinatorial metrics. accepted to *IEEE Int. Symp. on Inf. Theory*, Trondheim, 1994
2. H. D. Shapiro, Theoretical limitations on the efficient use of parallel memories. *IEEE Trans. Comput.* , vol. C-27, 421-428, May 1978

3. E. Belitskaya, V. Sidorenko and P. Stenström, Testing of memory with defects of fixed configuration. II Int. workshop *Algebraic and combinatorial coding theory*, Proc., Leningrad, USSR, 24-28, Sept. 1990
4. S. W. Golomb, Polyominos. Allen and Unwin, London, 1966
5. H. Wang, Proving theorems by pattern recognition - II. *Bell System Tech. J.*, 40, 1-41, 1961
6. R.Berger, The undecidability of the domino problem. *Mem. Amer. Math. Soc.*, 66, 1966
7. R. M. Robinson, Undecidability and nonperiodicity for tilings of the plane. *Invent. Math.*, 12, 177-209, 1971
8. H.A.G. Wijshoff and J.van Leuwen, Arbitrary versus periodic storage schemes and tessellations of the plane using one type of polyomino. *Inform. and Control*, vol. 62, 1-25, May 1984
9. D. Beauquier and M. Nivat, On translating one polyomino to tile the plane. *Discrete Comput. Geom.*, 6, 575-592, 1991
10. C. Radin, Tiling, periodicity, and crystals. *J. Math. Phys.*, 26(6), 1342-1344, June 1985
11. C. Radin and L. S. Schulman, Periodicity of classical ground states. *Phys. Review Letters*, 51(8), 621-622, Aug. 1983

Fast Infinite Response Filtering for Speech Processing

Irina E.Bocharova and Boris D.Kudryashov

St.-Petersburg Academy of Airspace Instrumentation,
Bolshaia Morskaia str., 67, St.-Petersburg,190000, Russia,
e-mail:liap@sovam.com

Abstract. The problem of reducing the number of multiplications in infinite response filtering is considered. We show how this operation may be decomposed on finite response filtrations and so-named half-convolutions. Examples of the efficient procedures for filters of low order are presented.

1 Introduction

Infinite response filtering (IRF) is widely applied in digital signal processing particularly in speech processing. In digital systems of this type a filter with infinite response is used to model the speech signal's short-term spectrum, or formant structure. The well-known linear prediction coding scheme [1] is based on IRF. The problem of reducing computational complexity in IRF is very important and becomes more important when one deals with the CELP standard.

An infinite response filtering is one of the main operations in the Code Excited Linear Prediction (CELP) speech compression system [2]. In this system a synthetic speech signal is obtained by passing the best (in the minimum squared error sense) excitation vector through an infinite response filter of order greater than or equal to 10. An excitation vector represents a sum of vectors from an adaptive code book (ACB) and a stochastic code book (SCB). The best excitation vector is searched through 128 code words of ACB and 1024 code words of SCB. Each vector is passed through the infinite response filter and the resulting signal is compared with original speech signal to make a decision concerning the best excitation vector. CELP standard implementation requires an astronomical number of IRFs since an exhaustive search through the code books for an excitation vector is implied.

In standard modifications to CELP designed to reduce computational complexity, the exhaustive search is replaced by a fast search with negligible loss in synthetic speech quality, but even in this case IRF takes the essential part of the running time of the CELP-like programs.

We consider the problem of minimizing the number of multiplications in filtering implementation. We illustrate our approach by some simple examples. We show that IRF may be reduced to computing finite response filtrations and so-named "half-convolutions". The generalization of our approach to filters of arbitrary order is straightforward.

2 Computational Algorithm

A finite impulse response digital filter is described by the expression

$$y_i = \sum_{j=0}^{n-1} x_j a_{i-j}, i = 0, 1, \ldots, n + m - 2, \tag{1}$$

where y_i, and x_j denote samples of output and input signals, a_k are the filter coefficients, m is the filter order and n is the input data block length. It is clear that the computation of (1) may be represented as multiplying two polynomials $a(z)$ and $x(z)$ with coefficients $a_i, i = 0, \ldots, m-1$ and $x_i, i = 0, \ldots, n-1$ respectively. In (1) we assume that $a_k = 0$ for $k < 0$. It is well-known that to multiply polynomials with the minimal number of multiplications one may use so-called "fast digital convolution" algorithms (FDCA) (see for example [3,4]). We remind what means "fast digital convolution" algorithm.

Example 1. Consider the following example. It is necessary to multiply two polynomials $a(z) = (a_0 + a_1 z)$ and $b(z) = (b_0 + b_1 z)$. The conventional algorithm requires 4 multiplications. One multiplication may be saved by presumming a_0, a_1 and b_0, b_1. In other words we calculate

$$c_0 = a_0 b_0, c_1' = (a_0 + a_1)(b_0 + b_1), c_2 = a_1 b_1.$$

and then

$$c_1 = c_1' - c_0 - c_2.$$

This method is called the Karatsuba algorithm [3]. It may be applied to the multiplication of two polynomials of order greater than 1. The next example shows how it may be done.

Example 2. It is necessary to multiply two third order polynomials $a(z) = a_0 + a_1 z + a_2 z^2 + a_3 z^3$ and $b(z) = b_0 + b_1 z + b_2 z^2 + b_3 z^3$. Introduce the following notation: $A_0(z) = a_0 + a_1 z$, $A_1(z) = a_2 z + a_3 z^2$, $B_0(z) = b_0 + b_1 z$,$B_1(z) = b_2 z + b_3 z^2$. Then the initial polynomials will have the form $a(z) = A_0(z) + A_1(z)z$ and $b(z) = B_0(z) + B_1(z)z$. Using the Karatsuba algorithm recurrently (for multiplying polynomials $a(z)$ and $b(z)$ and then for multiplying polynomials $A_i(z)$ and $B_j(z)$, $i, j = 0, 1$) we obtain the resulting polynomial $c(z)$. Thus 4-th order convolution is performed by employing 9 multiplications instead of 16.

In the general case the number of multiplications for large polynomial orders m is equal to $m^{log_2 3}$ instead of m^2 for the conventional algorithm.

An infinite impulse response filtering is determined by the expression

$$y_i = \sum_{j=0}^{m-1} a_j y_{i-j} + x_i, i = 1, 2, \ldots. \tag{2}$$

and in this case the FDCA approach can not be employed directly.

We show that FDCA may be applied to computing (2) if one computes a block of t samples of the output signal.

Example 3. Let m be equal to 12 and t be equal to 4. Then four consecutive samples of the filter output signal have the form

$$\begin{aligned} y_{i+1} &= a_1 y_i + a_2 y_{i-1} + a_3 y_{i-2} + a_4 y_{i-3} + a_5 y_{i-4} + a_6 y_{i-5} + a_7 y_{i-6} \\ &\quad + a_8 y_{i-7} + a_9 y_{i-8} + a_{10} y_{i-9} + a_{11} y_{i-10} + a_{12} y_{i-11} + x_i \\ y_{i+2} &= a_1 y_{i+1} + a_2 y_i + a_3 y_{i-1} + a_4 y_{i-2} + a_5 y_{i-3} + a_6 y_{i-4} + a_7 y_{i-5} \\ &\quad + a_8 y_{i-6} + a_9 y_{i-7} + a_{10} y_{i-8} + a_{11} y_{i-9} + a_{12} y_{i-10} + x_{i+1} \\ y_{i+3} &= a_1 y_{i+2} + a_2 y_{i+1} + a_3 y_i + a_4 y_{i-1} + a_5 y_{i-2} + a_6 y_{i-3} + a_7 y_{i-4} \\ &\quad + a_8 y_{i-5} + a_9 y_{i-6} + a_{10} y_{i-7} + a_{11} y_{i-8} + a_{12} y_{i-9} + x_{i+2} \\ y_{i+4} &= a_1 y_{i+3} + a_2 y_{i+2} + a_3 y_{i+1} + a_4 y_i + a_5 y_{i-1} + a_6 y_{i-2} + a_7 y_{i-3} \\ &\quad + a_8 y_{i-4} + a_9 y_{i-5} + a_{10} y_{i-6} + a_{11} y_{i-7} + a_{12} y_{i-8} + x_{i+4} \end{aligned} \tag{3}$$

The problem of calculating samples $y_{i+k}, k = 1, \ldots t$ may be reduced to the problem of calculating three fourth order convolutions. They may be written in polynomial form as follows

$$\begin{gathered} (a_5 + a_6 z + a_7 z^2 + a_8 z^3)\,(y_{i-4} + y_{i-3} z + y_{i-2} z^2 + y_{i-1} z^3) \\ (a_9 + a_{10} z + a_{11} z^2 + a_{12} z^3)\,(y_{i-8} + y_{i-7} z + y_{i-6} z^2 + y_{i-5} z^3) \\ (a_1 + a_2 z + a_3 z^2 + a_4 z^3)\,(y_{i-3} + y_{i-2} z + y_{i-1} z^2 + y_i z^3) \end{gathered}$$

The first convolution gives the items

$$\begin{aligned} &a_5 y_{i-4} \\ &a_5 y_{i-3} + a_6 y_{i-4} \\ &a_5 y_{i-2} + a_6 y_{i-3} + a_7 y_{i-4} \\ &a_5 y_{i-1} + a_6 y_{i-2} + a_7 y_{i-3} + a_8 y_{i-4} \end{aligned}$$

which are used at the current step (for computing $y_{i+1}, \ldots, y_{i+4}$) and the items

$$\begin{aligned} a_6 y_{i-1} + a_7 y_{i-2} + a_8 y_{i-3} \\ a_7 y_{i-1} + a_8 y_{i-2} \\ a_8 y_{i-1} \end{aligned}$$

which will be used at the next step (for computing $y_{i+5}, \ldots, y_{i+8}$). The second convolution generates the terms

$$\begin{aligned} &a_9 y_{i-8} \\ &a_9 y_{i-7} + a_{10} y_{i-8} \\ &a_9 y_{i-6} + a_{10} y_{i-7} + a_{11} y_{i-8} \\ &a_9 y_{i-5} + a_{10} y_{i-6} + a_{11} y_{i-7} + a_{12} y_{i-8} \end{aligned}$$

for the current step and the terms

$$\begin{aligned} a_{10} y_{i-5} + a_{11} y_{i-6} + a_{12} y_{i-7} \\ a_{11} y_{i-5} + a_{12} y_{i-6} \\ a_{12} y_{i-5} \end{aligned}$$

for the next step. The third convolution of two third-order polynomials has half terms known from the previous step. More exactly the terms

$$\begin{aligned}&a_1 y_{i-3}\\&a_1 y_{i-2} + a_2 y_{i-3}\\&a_1 y_{i-1} + a_2 y_{i-2} + a_3 y_{i-3}\end{aligned}$$

were obtained when $y_{i-3}, \ldots, y_i$ have been calculated. The items

$$\begin{aligned}a_1 y_i + a_2 y_{i-1} + a_3 y_{i-2} + a_4 y_{i-3}\\a_2 y_i + a_3 y_{i-1} + a_4 y_{i-2}\\a_3 y_i + a_4 y_{i-1}\\a_4 y_i\end{aligned}$$

are computed at the current step. The coefficients

$$\begin{aligned}&a_1 y_{i+1}\\&a_1 y_{i+2} + a_2 y_{i+1}\\&a_1 y_{i+3} + a_2 y_{i+2} + a_3 y_{i+1}\end{aligned}$$

are calculated directly at the current step.

Therefore to compute infinite impulse response (3) we calculate three finite impulse responses. One of them is computed under the assumption that 3 terms of the convolution are already known.

3 "Half-Convolution"

Convolution with some coefficients known in advance we shall call "half-convolution".

Example 4. Consider convolution of order 2×2 or the multiplication of two first order polynomials

$$(a_0 + a_1 z)(b_0 + b_1 z) = c_0 + c_1 z + c_2 z^2$$

with one coefficient of the result (c_2) known in advance. Then if one uses the conventional algorithm three multiplications are required to compute coefficients c_0 and c_1. However it is rather evident that these coefficients may be obtained as follows:

$$c_0 = a_0 b_0, c_1 = (a_0 + a_1)(b_0 + b_1) - c_0 - c_2$$

and in this case only two multiplications are required.

In the general case the method of obtaining "half-convolution" coefficients may be developed using approach analogous to well-known Cook-Toom algorithm [3].

Example 5. Consider the 3×3 convolution

$$(a_0 + a_1 z + a_2 z^2)(b_0 + b_1 z + b_2 z^2) = c_0 + c_1 z + c_2 z^2 + c_3 z^3 + c_4 z^4$$

where c_3 and c_4 are already known. The rest of the coefficients may be computed as follows

$$\begin{aligned}
c_0 &= a_0 b_0,\\
c_1' &= (a_0 + a_1 + a_2)(b_0 + b_1 + b_2),\\
c_2' &= (a_0 - a_1 + a_2)(b_0 - b_1 + b_2),\\
c_1 &= (c_1' - c_2')/2 - c_3,\\
c_2 &= (c_1' + c_2')/2 - c_0 - c_4.
\end{aligned}$$

Example 6. Consider 4×4 convolution of two third-order polynomials with coefficients $a_0, \ldots, a_3$ and $b_0, \ldots, b_3$ if coefficients c_4, c_5, c_6 of the result are known. Then the other convolution coefficients $c_0, \ldots c_3$ may be calculated by the following algorithm

$$\begin{aligned}
c_0 &= a_0 b_0,\\
c_1 &= a_0 b_1 + a_1 b_0,\\
c_1' &= (a_0 + a_1 + a_2 + a_3)(b_0 + b_1 + b_2 + b_3),\\
c_2' &= (a_0 - a_1 + a_2 - a_3)(b_0 - b_1 + b_2 - b_3),\\
c_2 &= (c_1' + c_2')/2 - c_0 - c_4 - c_6,\\
c_3 &= (c_1' - c_2')/2 - c_1 - c_5.
\end{aligned}$$

It is easy to see that to perform "half-convolution" only 5 multiplications are necessary instead of 10 for the conventional technique or 9 for the Karatsuba scheme.

The performance of some short half-convolutions is given in the Table 1.

Table 1. Performance of short half-convolutions.

m	n	The number of multiplications for full convolutions	The number of multiplications for half-convolutions
2	2	3	2
3	3	6(5)	3
4	4	9(7)	5(4)
4	5	11(8)	7(5)

Note that for 4 × 4 and 4 × 5 convolutions and half-convolutions the number of multiplications may be reduced (see, for example, [3]) to values pointed out in the parenthesis in Table 1. However the corresponding algorithms require the premultiplication of filter coefficients by fractions such as 1/6, or in other words, they require division. This problem may be easily overcome if a filter is predetermined (filter coefficients are fixed). In CELP-based systems filter coefficients vary for successive frames. So, we prefer suboptimal (in the sense of minimizing the number of multiplications) algorithms using only division by 2, which is implemented as a shift.

4 Conclusion

We investigated the block lengths $t = 4, 5, 8$ for the filter order $m = 12$. The results obtained are shown in Table 2. Here we denote the convolution of two sequences of length k and l Пas $(k \times l)$.

Table 2. Number of multiplications for block lengths $t = 4, 5, 8$ for 12-th order IRF.

t	The number of convolutions per block	The number of half-convolutions per block	The number of multiplications per sample
4	2 (4×4)	1 (4×4)	29/4
5	2 (4×5)	1 (4×5)	38/5
8	1 (8×8)	2 (4×4)	49/8

It follows from Table 2 that for $t = 4, m = 12$ filtering may be reduced to calculating two convolutions of fourth order and one convolution of fourth order with 3 coefficients known in advance. In this case only 29/4 multiplications per sample instead of 12 are required. For $t = 8$ the number of multiplications is reduced almost twice. Note that although the number of additions for the proposed algorithm is greater than for the conventional one, the total number of operations (additions+multiplications) remains approximately the same.

The running time of the filtering programs using the methods described here was reduced by 30-40 percent for an IBM PC 486 computer in comparison with conventional filtering.

References

1. A.V.Oppenheim, *Applications of Digital Signal Processing.* Prentice-Hall, Inc., Englewood Cliffs, New Jersey,1978.
2. J.P.Campbell, V.C.Welch and T.E.Tremain,*CELP Documentation Version 3.2.* U.S.Dod, Fort Mead, MD, September, 1990
3. R.E. Blahut, *Fast algorithms for digital signal processing.* Addison-Wessley, Reading, Mass.,1985.
4. A.V.Aho, J.E.Hopcroft, J.D.Ullman, *The Design and Analysis of Computer Algorithms.* Addison-Wessley, Reading, Mass., 1976.

On Trellis Codes for Linear Predictive Speech Compression

Irina E. Bocharova, Victor D. Kolesnik, Victor Yu. Krachkovsky,
Boris D. Kudryashov, Eugeny P. Ovsjannikov, Boris K. Troyanovsky

St.-Petersburg Academy of Aerospace Instrumentation,
Bolshaya Morskaya str., 67, St.-Petersburg, 190000, Russia.
e-mail: liap@sovam.com

Abstract. We consider the problem of efficient codebook search for the code excited linear predictive (CELP) speech coders. To diminish the search complexity a trellis codebook is proposed and decoding algorithms based on the Viterbi algorithm are investigated. Numerical results of the computer simulation are presented.

1 Introduction

One of the best known methods of low-rate high-quality speech compression is CELP (Code Excited Linear Prediction), originally proposed by B.S.Atal [1, 2]. The CELP-based approach was used for developing US Federal Standard 1016 for speech quantization at 4800 bit/sec [3]. The main difficulty with CELP lies in the complexity of the search performed by the stochastic speech analyzer (SSA). The search problem can be described as follows.

The input speech is partitioned into subframes of length N, $N \approx 60 - 80$. To obtain an input vector $\mathbf{v} = (v_1, \dots, v_N)$, $-\infty < v_i < +\infty$, of SSA the periodic (pitch) and previous subframe memory (ringing) components are subtracted from the speech subframe, preliminary filtered by a perceptual filter. Let $\mathcal{C} = \{\mathbf{c}_1, \mathbf{c}_2, \dots, \mathbf{c}_M\}$ be a codebook consisting of M words, $M = 2^\nu, \nu \approx 8 - 10$, of length N. We will denote by $1/A(z)$ the SSA synthesis filter, where $A(z) = 1 - a_1 \cdot z - \dots - a_p \cdot z^p$ is a polynomial of degree p, computed by the well known filter synthesis techniques (see, for example, [4]). For CELP-based algorithms the values $p = 10 - 12$ are recommended. The best code excitation vector $\mathbf{c}_0 \in \mathcal{C}$ and the best gain g_0, $-\infty < g_0 < \infty$, are searched by using an analysis-by-synthesis approach. By this approach the input vector $\mathbf{v}$ is approximated by the vector $\mathbf{w}$, obtained as an output, scaled by g_0, of the filter $1/A(z)$ excited by the vector $\mathbf{c}_0 \in \mathcal{C}$.

This search problem can be formulated in a matrix form. Let $H(z) = 1/A(z)$ $= h_0 + h_1 \cdot z + h_2 \cdot z^2 + \dots + h_{N-1} \cdot z^{N-1} + \dots$ define the z-transform of the impulse response of SSA synthesis filter. We define a triangular Toeplitz matrix

$$\mathbf{H} = \begin{bmatrix} h_0 \; h_1 \; h_2 \dots h_{N-1} \\ 0 \;\; h_0 \; h_1 \dots h_{N-2} \\ \dots \\ 0 \;\; 0 \;\; 0 \dots \;\; h_0 \end{bmatrix} .$$

Then $\mathbf{w} = g \cdot \mathbf{c} \cdot \mathbf{H}$ represents the output of the filter $1/A(z)$ excited by $\mathbf{c}$ and the problem is to find the excitation vector $\mathbf{c}_0 \in \mathcal{C}$, and the gain g_0, $-\infty < g_0 < \infty$ that jointly minimize the squared euclidean distance between the original and synthesized speech vectors

$$(\mathbf{c}_0, g_0) = \arg\min_{\mathbf{c}\in\mathcal{C}, g} d_E^2(\mathbf{v}, g \cdot \mathbf{c} \cdot \mathbf{H}) \tag{1}$$

where $d_E(\mathbf{x}, \mathbf{y})$ is the euclidean distance between vectors $\mathbf{x}$ and $\mathbf{y}$ and

$$\begin{aligned} d_E^2(\mathbf{v}, g \cdot \mathbf{c} \cdot \mathbf{H}) &= \|\mathbf{v} - g \cdot \mathbf{c} \cdot \mathbf{H}\|^2 \\ &= (\mathbf{v}, \mathbf{v}) - 2g \cdot (\mathbf{v}, \mathbf{c} \cdot \mathbf{H}) + g^2 \cdot (\mathbf{c} \cdot \mathbf{H}, \mathbf{c} \cdot \mathbf{H}). \end{aligned} \tag{2}$$

Given a codeword $\mathbf{c} \in \mathcal{C}$, we can find a minimum in (1) from the equation

$$\frac{\partial}{\partial g} d_E^2(\mathbf{v}, g \cdot \mathbf{c} \cdot \mathbf{H}) = 0.$$

Therefore

$$g = \frac{(\mathbf{v}, \mathbf{c} \cdot \mathbf{H})}{(\mathbf{c} \cdot \mathbf{H}, \mathbf{c} \cdot \mathbf{H})}. \tag{3}$$

By substituting (3) into (1) and (2) we can find the optimal codeword as

$$\mathbf{c}_0 = \arg\max_{\mathbf{c}\in\mathcal{C}} \mu(\mathbf{v}, \mathbf{c} \cdot \mathbf{H}), \qquad \mu(\mathbf{v}, \mathbf{c} \cdot \mathbf{H}) = \frac{(\mathbf{v}, \mathbf{c} \cdot \mathbf{H})^2}{(\mathbf{c} \cdot \mathbf{H}, \mathbf{c} \cdot \mathbf{H})}, \tag{4}$$

where μ is called the match function. In order to perform the search over a codebook of size 2^{10} with codeword length 60, about $2 \cdot 10^6$ multiplications are required. The extremely high complexity of the search is a major difficulty with a real-time implementation of CELP on a digital signal processor (DSP).

In Section 2 we consider a construction of a codebook from a trellis code. As it will be shown in Section 3, by using the trellis codebook and quasi-optimal search procedures based on Viterbi decoding we can considerably reduce the search complexity for CELP-based algorithms. The speech compression characteristics are compared in Section 4.

2 Trellis Code Book

To avoid an exhaustive search over a codebook of large size, we introduce a codebook constructed from a given trellis code. For simplicity only time-invariant regular trellis codes are considered. Let $\mathcal{T} = \mathcal{T}(\mathcal{S}, \mathcal{E})$ be a trellis defined by a set of states $\mathcal{S}$ of size $S = |\mathcal{S}|$, set of edges $\mathcal{E}$, and levels $i = 1, 2, \ldots$. Each edge $e \in \mathcal{E}$ begins at some state $\beta(e) \in \mathcal{S}$ of level $i-1$ and terminates at a state $\tau(e) \in \mathcal{S}$ of the next level i. For each state s we define by $E_\tau(s) \subseteq \mathcal{E}$, the set of edges terminating at s and by $E_\beta(s) \subseteq \mathcal{E}$ the set of edges emerging from s.

We call $\mathcal{T}$ a regular trellis if for each $s \in \mathcal{S}$, $E_\beta(s)$ does not depend on the trellis level and $|E_\beta(s)| = 2^k$ for some integer k. A sequence of edges

$\gamma = (e_1, e_2, \ldots, e_l)$ is called a path of length l beginning at a state s, if $\beta(e_1) = s$ and $\beta(e_i) = \tau(e_{i-1})$ for $2 \le i \le l$.

For a given trellis T we can construct a trellis code by labelling each edge $e \in \mathcal{E}$ by a subblock of n symbols $\mathbf{u} = \mathbf{u}(e) \in \mathcal{U}$, where $\mathcal{U}$ is a set of all different edge labellings. The rate of the trellis code is equal to $R = k/n$. We suppose that $\mathcal{U} \subseteq \mathcal{Q}^n$, where $\mathcal{Q}$ is a code alphabet. A binary alphabet $\mathcal{Q}_b = \{-1, +1\}$ and a ternary alphabet $\mathcal{Q}_t = \{-1, 0, +1\}$ are of interest. Each path $\gamma = (e_1, e_2, \ldots, e_l)$ of length l in a trellis generates a codeword $\mathbf{c} = (\mathbf{c}_1, \mathbf{c}_2, \ldots, \mathbf{c}_l) \in \mathcal{C_T}$ of length $N = l \cdot n$, where $\mathbf{c}_i = \mathbf{u}(e_i)$ and there exists exactly $M = S \cdot 2^{k \cdot l}$ such different paths. We define a trellis codebook $\mathcal{C_T}$ as a block code of length N and size M with the codewords, presented by all different paths on a trellis.

The simplest way to construct a trellis codebook over the binary alphabet is to use a binary linear trellis code $\mathcal{A}$ over the alphabet $\mathcal{Q}_b' = \{0, 1\}$, replacing $0 \mapsto 1$, $1 \mapsto -1$. As it follows from a computer simulation the code $\mathcal{A}$ should satisfy the following requirements:

- $\mathcal{A}$ should be a "good" covering code.
- The number of states S should be sufficiently large.
- $\mathcal{A}$ should have a particular power spectra (the spectral maximums should be disposed on such frequencies which yield the minimal noise for the synthesized speech).

A fragment of the trellis codebook $\mathcal{C_T}$ obtained from the well known (5,7) (defined by the octal notation of its generator polynomials) convolutional code is depicted in Fig.1.

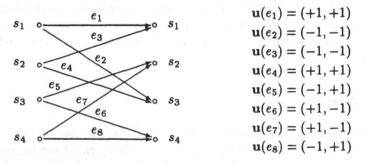

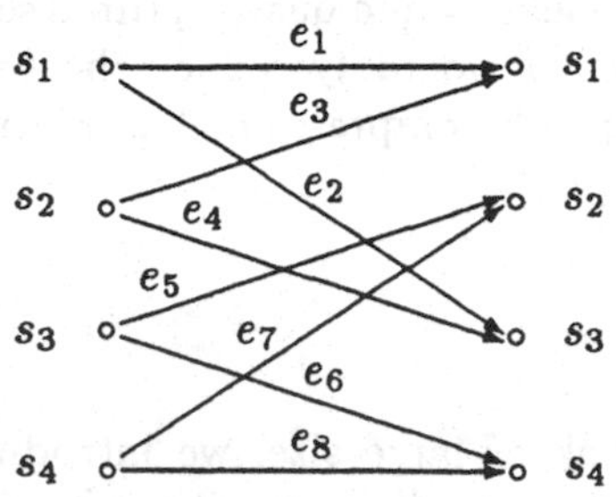

$$\mathbf{u}(e_1) = (+1, +1)$$
$$\mathbf{u}(e_2) = (-1, -1)$$
$$\mathbf{u}(e_3) = (-1, -1)$$
$$\mathbf{u}(e_4) = (+1, +1)$$
$$\mathbf{u}(e_5) = (-1, +1)$$
$$\mathbf{u}(e_6) = (+1, -1)$$
$$\mathbf{u}(e_7) = (+1, -1)$$
$$\mathbf{u}(e_8) = (-1, +1)$$

Fig. 1. A fragment of the trellis codebook, obtained from the (5,7)-code.

Binary codebooks, however, do not present the best choice for speech compression. A computer simulation shows that the better quality of synthesized speech is obtained by using a ternary codebook or even a codebook with more complex labelling of edges.

3 Decoding Algoritms

3.1 Preliminary Definitions

Upon receiving the input vector $\mathbf{v}$, an optimal decoder finds the word $\mathbf{c}_0 \in \mathcal{C}_T$ that satisfies condition (4). Since μ is not an additive function, we cannot use it as a metric function for the Viterbi algorithm in a straightforward way. We consider several quasi-optimal procedures based on the Viterbi algorithm. All algorithms process the SSA input vector $\mathbf{v}$ partitioned into l subblocks $\mathbf{v} = (\mathbf{v}_1, \mathbf{v}_2, \ldots, \mathbf{v}_l)$. In the first algorithm the gain estimate $\hat{g}_0$ is computed before the search of $\mathbf{c}_0$. In the other algorithms the gain is estimated after the search of $\mathbf{c}_0$ is completed.

Let $\gamma = (e_1, e_2, \ldots, e_l)$ be a path in the trellis, generating a word $\mathbf{c} \in \mathcal{C}_T$. For $1 \le i \le l$ we define by $\gamma^{(i)} = (e_1, e_2, \ldots, e_i)$ a path from the first i edges of γ, and by $\mathbf{c}^{(i)} = (\mathbf{c}_1, \mathbf{c}_2, \ldots, \mathbf{c}_i)$ a subword of $\mathbf{c}$, generated by this path. We also define the following submatrices of $\mathbf{H}$

$$\mathbf{H}_i = \begin{bmatrix} h_0 & h_1 & h_2 & \ldots & h_{i\cdot n-1} \\ 0 & h_0 & h_1 & \ldots & h_{i\cdot n-2} \\ & & \cdots & & \\ 0 & 0 & 0 & \ldots & h_0 \end{bmatrix}, \quad \mathbf{B}_i = \begin{bmatrix} h_{(i-1)\cdot n} & h_{i\cdot n+1} & \cdots & h_{i\cdot n-1} \\ h_{(i-1)\cdot n-1} & h_{i\cdot n} & \cdots & h_{i\cdot n-2} \\ \cdots & \cdots & \cdots & \cdots \\ h_0 & h_1 & \cdots & h_{n-1} \end{bmatrix},$$

for $1 \le i \le l$, $\mathbf{H}_l = \mathbf{H}$. Since $\mathbf{H}_i$ is a triangular Toeplitz matrix it may be presented as

$$\mathbf{H}_i = \left[\begin{array}{c|c} \mathbf{H}_{i-1} & \mathbf{B}_i \\ \mathbf{0} & \mathbf{H}_1 \end{array}\right], \quad 2 \le i \le l, \tag{5}$$

where $\mathbf{0}$ denotes the zero matrix of size $n \times n$. From (5) we obtain the following result.

Proposition 1 *Let* $\mathbf{w}^{(i)} = (\mathbf{w}_1, \mathbf{w}_2, \ldots, \mathbf{w}_i) = \mathbf{c}^{(i)} \cdot \mathbf{H}_i$ *be an SSA synthesis filter response to the excitation* $\mathbf{c}^{(i)}$. *Then* $\mathbf{w}^{(i)}$ *can be constructed by appending*

$$\mathbf{w}^{(i)} = \mid \mathbf{w}^{(i-1)} \mid \mathbf{w}_i \mid, \tag{6}$$

where

$$\mathbf{w}_i = \mathbf{w}'_i + \mathbf{w}''_i, \qquad \mathbf{w}'_i = \mathbf{c}_i \cdot \mathbf{H}_1 \quad \text{and} \quad \mathbf{w}''_i = \mathbf{c}^{(i-1)} \cdot \mathbf{B}_i. \tag{7}$$

The subblock $\mathbf{w}'_i$ in (7) is a response of the SSA synthesis filter excited by the subblock $\mathbf{c}_i$ from a zero initial state, while $\mathbf{w}''_i$ is a response of this filter for a zero input starting from the initial filter state reached for the previous excitation $\mathbf{c}^{(i-1)}$. The expression (7) may be used in the decoding algorithm to compute the subblocks $\mathbf{w}_i$. However, the recursive calculation of $\mathbf{w}''_i$ by autoregressive filtration [4] usually requires less add-multiply operations than the matrix calculation of $\mathbf{w}''_i$ by the expression (7).

Let $s_j = \tau(e_i^{(i)})$ be the terminating state of the path $\gamma^{(i)}$ that generates the subword $\mathbf{c}^{(i)}$ and let us define

$$\kappa_{i,j} = (\mathbf{v}^{(i)}, \mathbf{w}^{(i)}), \quad \varepsilon_{i,j} = (\mathbf{w}^{(i)}, \mathbf{w}^{(i)}). \tag{8}$$

We call $\kappa_{i,j}$ a partial correlation and $\varepsilon_{i,j}$ a partial energy. Since these functions are additive functions of the vectors $\mathbf{v}^{(i)}$, $\mathbf{w}^{(i)}$, they can be used as accumulated values for algorithms based on the Viterbi decoding. However, after the substitution $\mathbf{w}^{(i)} = \mathbf{c}^{(i)} \cdot \mathbf{H}_i$, the functions κ_i and ε_i will no longer be additive functions of the vectors $\mathbf{v}^{(i)}$ and $\mathbf{c}^{(i)}$. This could result in some additional decrease in the Viterbi decoding performance with respect to an exhaustive search strategy for (1).

3.2 Euclidean Distance Decoding

We can apply the Viterbi algorithm to find the quasi-optimal solution of (1) if we replace the unknown gain in (2) by the estimate $\hat{g}$, obtained from the input vector $\mathbf{v}$ and the impulse response $H(z)$ (for an application of Viterbi algorithm for the euclidean distance decoding with higher compression rate see [5]). The best codevector $\mathbf{c}_{0,ed} \in \mathcal{C}_T$ is searched by successively finding for each state $s_j \in \mathcal{S}$ of the level $i = 2, 3, \ldots, l$ such a path $\gamma^{(i)} = (e_1^{(i)}, e_2^{(i)}, \ldots, e_i^{(i)})$, $\tau(e_i^{(i)}) = s_j$, that minimizes the squared euclidean distance

$$\begin{aligned} \delta_{i,j} &= d_E^2(\mathbf{v}^{(i)}, \hat{g} \cdot \mathbf{c}^{(i)} \cdot \mathbf{H}_i) = \|\mathbf{v}^{(i)} - \hat{g} \cdot \mathbf{c}^{(i)} \cdot \mathbf{H}_i\|^2 \\ &= \rho_i - 2 \cdot \hat{g} \cdot \kappa_{i,j} + \hat{g}^2 \cdot \varepsilon_{i,j}, \end{aligned} \tag{9}$$

where $\rho_i = (\mathbf{v}^{(i)}, \mathbf{v}^{(i)})$, and $\mathbf{c}^{(i)}$ is a subvector, generated by the path $\gamma^{(i)}$. For each state s_j of the level i the values $\delta_{i,j}$ as well as the SSA synthesis filter states should be stored. At the last level l the best state $s_j \in \mathcal{S}$ providing a minimum for $\delta_{l,j}$ is searched and the codevector $\mathbf{c}_{0,ed}$ is finally generated.

3.3 Match Function Decoding

Following (8) we can define a partial match function of (4) for a subvector $\mathbf{c}^{(i)} = (\mathbf{c}_1^{(i)}, \mathbf{c}_2^{(i)}, \ldots, \mathbf{c}_l^{(i)})$ generated by some path $\gamma^{(i)} = (e_1^{(i)}, e_2^{(i)}, \ldots, e_i^{(i)})$, with a terminating state $s_j = \tau(e_i^{(i)})$ as

$$\mu_{i,j} = \frac{(\mathbf{v}^{(i)}, \mathbf{c}^{(i)} \cdot \mathbf{H}_i)^2}{(\mathbf{c}^{(i)} \cdot \mathbf{H}_i, \mathbf{c}^{(i)} \cdot \mathbf{H}_i)} = \frac{\kappa_{i,j}^2}{\varepsilon_{i,j}}, \tag{10}$$

where $\kappa_{i,j}$ and $\varepsilon_{i,j}$ are defined by (8). We search the best codevector $\mathbf{c}_{0,mf} \in \mathcal{C}_T$ by successively finding for each state $s_j \in \mathcal{S}$ of the level $i = 2, 3, \ldots, l$ such a path $\gamma^{(i)}$ terminating at s_j that gives a maximum for $\mu_{i,j}$ in (10). To control the search, the values $\kappa_{i,j}$, $\varepsilon_{i,j}$ as well as the SSA synthesis filter states should be stored for each state s_j of the level i. The search stops after selecting the state s_{j_0} of the level l that provides the maximal value of μ_{l,j_0}. The surviving path leading to this state generates the resulting excitation vector $\mathbf{v}_{0,mf}$. The corresponding gain value can be computed as $g_{0,mf} = \kappa_{l,j_0}/\varepsilon_{l,j_0}$.

3.4 Correlation Decoding

The algorithm of the previous subsection can be simplified if on each step of the procedure only the denominator in the expression (10) is calculated. Thus for each state $s_j \in S$ on a level i a path with the maximal value of $| \kappa_{i,j} |$ survives. This simplification of metrics could be a significant advantage for a fixed point implementation.

4 Simulation Results

The Trellis Excited Linear Predictive (TELP) speech compression algorithm was simulated on an IBM-486, 66 MHz computer. The compression bitrate was varied in the range 2400-8000 bit/s for different trellis codebooks, subframe lengths and other parameters. The results of listening tests show a good quality of speech at 4800 bits/s at the normalized MSE for perceptual speech in the range 0.1-0.25. The time of speech compression was about 2 sec. for 1 sec. of speech. The results of simulation for 12 files with voices from different male, female and children speakers are presented in Table 1

TABLE I

Normalized MSE for the perceptual speech.

File number	Euclidean distance decoding	Match function decoding	Correlation decoding
1	0.229	0.221	0.221
2	0.196	0.192	0.159
3	0.280	0.278	0.300
4	0.220	0.197	0.202
5	0.220	0.212	0.240
6	0.180	0.147	0.160
7	0.200	0.170	0.186
8	0.330	0.300	0.340
9	0.298	0.269	0.298
10	0.320	0.295	0.288
11	0.211	0.168	0.186
12	0.230	0.186	0.220

The algorithm was also realized in real time on two DSP's with fixed point arithmetic, 25 mHz clock frequency, 1K words RAM, 4K words ROM for the bitrate 2400 bit/s and 10 mHz clock frequency, 512 words RAM, 4K words ROM for the bitrate 4800 bit/s. The quality for both cases was good enough for telephone speech transmission.

References

1. B.S.Atal, "Predictive Coding of Speech at Low Bit Rates," *IEEE Trans.*, vol. Com-30, No.4, pp.600-614, Apr.1982.
2. M.R.Schroeder and B.S.Atal, "Code Excited Linear Prediction (CELP): High quality Speech at Very Low Bit Rates," *International Conference on Acoustics, Speech and Signal Processing, Proceedings* , pp.937-940, 1982.
3. J.P.Campbell, V.C.Welch and T.E.Tremain, *CELP Documentation Version 3.2.* U.S.Dod,Fort Mead, MD,September,1990.
4. A.V.Oppenheim and R.W.Schafer, *Digital Signal Processing.* Englewood Cliffs, N.J.: Prentice-Hall, 1975.
5. M.W.Marcellin, T.S.Fisher and J.D.Gibson,
"Predictive Trellis Coded Quantization of Speech," *IEEE Trans.*, vol. ASSP-38, No.1, pp.46-55, Jan. 1990.

Lecture Notes in Computer Science

For information about Vols. 1–751
please contact your bookseller or Springer-Verlag

Vol. 752: T. W. Finin, C. K. Nicholas, Y. Yesha (Eds.), Information and Knowledge Management. Proceedings, 1992. VII, 142 pages. 1993.

Vol. 753: L. J. Bass, J. Gornostaev, C. Unger (Eds.), Human-Computer Interaction. Proceedings, 1993. X, 388 pages. 1993.

Vol. 754: H. D. Pfeiffer, T. E. Nagle (Eds.), Conceptual Structures: Theory and Implementation. Proceedings, 1992. IX, 327 pages. 1993. (Subseries LNAI).

Vol. 755: B. Möller, H. Partsch, S. Schuman (Eds.), Formal Program Development. Proceedings. VII, 371 pages. 1993.

Vol. 756: J. Pieprzyk, B. Sadeghiyan, Design of Hashing Algorithms. XV, 194 pages. 1993.

Vol. 757: U. Banerjee, D. Gelernter, A. Nicolau, D. Padua (Eds.), Languages and Compilers for Parallel Computing. Proceedings, 1992. X, 576 pages. 1993.

Vol. 758: M. Teillaud, Towards Dynamic Randomized Algorithms in Computational Geometry. IX, 157 pages. 1993.

Vol. 759: N. R. Adam, B. K. Bhargava (Eds.), Advanced Database Systems. XV, 451 pages. 1993.

Vol. 760: S. Ceri, K. Tanaka, S. Tsur (Eds.), Deductive and Object-Oriented Databases. Proceedings, 1993. XII, 488 pages. 1993.

Vol. 761: R. K. Shyamasundar (Ed.), Foundations of Software Technology and Theoretical Computer Science. Proceedings, 1993. XIV, 456 pages. 1993.

Vol. 762: K. W. Ng, P. Raghavan, N. V. Balasubramanian, F. Y. L. Chin (Eds.), Algorithms and Computation. Proceedings, 1993. XIII, 542 pages. 1993.

Vol. 763: F. Pichler, R. Moreno Díaz (Eds.), Computer Aided Systems Theory – EUROCAST '93. Proceedings, 1993. IX, 451 pages. 1994.

Vol. 764: G. Wagner, Vivid Logic. XII, 148 pages. 1994. (Subseries LNAI).

Vol. 765: T. Helleseth (Ed.), Advances in Cryptology – EUROCRYPT '93. Proceedings, 1993. X, 467 pages. 1994.

Vol. 766: P. R. Van Loocke, The Dynamics of Concepts. XI, 340 pages. 1994. (Subseries LNAI).

Vol. 767: M. Gogolla, An Extended Entity-Relationship Model. X, 136 pages. 1994.

Vol. 768: U. Banerjee, D. Gelernter, A. Nicolau, D. Padua (Eds.), Languages and Compilers for Parallel Computing. Proceedings, 1993. XI, 655 pages. 1994.

Vol. 769: J. L. Nazareth, The Newton-Cauchy Framework. XII, 101 pages. 1994.

Vol. 770: P. Haddawy (Representing Plans Under Uncertainty. X, 129 pages. 1994. (Subseries LNAI).

Vol. 771: G. Tomas, C. W. Ueberhuber, Visualization of Scientific Parallel Programs. XI, 310 pages. 1994.

Vol. 772: B. C. Warboys (Ed.),Software Process Technology. Proceedings, 1994. IX, 275 pages. 1994.

Vol. 773: D. R. Stinson (Ed.), Advances in Cryptology – CRYPTO '93. Proceedings, 1993. X, 492 pages. 1994.

Vol. 774: M. Banâtre, P. A. Lee (Eds.), Hardware and Software Architectures for Fault Tolerance. XIII, 311 pages. 1994.

Vol. 775: P. Enjalbert, E. W. Mayr, K. W. Wagner (Eds.), STACS 94. Proceedings, 1994. XIV, 782 pages. 1994.

Vol. 776: H. J. Schneider, H. Ehrig (Eds.), Graph Transformations in Computer Science. Proceedings, 1993. VIII, 395 pages. 1994.

Vol. 777: K. von Luck, H. Marburger (Eds.), Management and Processing of Complex Data Structures. Proceedings, 1994. VII, 220 pages. 1994.

Vol. 778: M. Bonuccelli, P. Crescenzi, R. Petreschi (Eds.), Algorithms and Complexity. Proceedings, 1994. VIII, 222 pages. 1994.

Vol. 779: M. Jarke, J. Bubenko, K. Jeffery (Eds.), Advances in Database Technology — EDBT '94. Proceedings, 1994. XII, 406 pages. 1994.

Vol. 780: J. J. Joyce, C.-J. H. Seger (Eds.), Higher Order Logic Theorem Proving and Its Applications. Proceedings, 1993. X, 518 pages. 1994.

Vol. 781: G. Cohen, S. Litsyn, A. Lobstein, G. Zémor (Eds.), Algebraic Coding. Proceedings, 1993. XII, 326 pages. 1994.

Vol. 782: J. Gutknecht (Ed.), Programming Languages and System Architectures. Proceedings, 1994. X, 344 pages. 1994.

Vol. 783: C. G. Günther (Ed.), Mobile Communications. Proceedings, 1994. XVI, 564 pages. 1994.

Vol. 784: F. Bergadano, L. De Raedt (Eds.), Machine Learning: ECML-94. Proceedings, 1994. XI, 439 pages. 1994. (Subseries LNAI).

Vol. 785: H. Ehrig, F. Orejas (Eds.), Recent Trends in Data Type Specification. Proceedings, 1992. VIII, 350 pages. 1994.

Vol. 786: P. A. Fritzson (Ed.), Compiler Construction. Proceedings, 1994. XI, 451 pages. 1994.

Vol. 787: S. Tison (Ed.), Trees in Algebra and Programming – CAAP '94. Proceedings, 1994. X, 351 pages. 1994.

Vol. 788: D. Sannella (Ed.), Programming Languages and Systems – ESOP '94. Proceedings, 1994. VIII, 516 pages. 1994.

Vol. 789: M. Hagiya, J. C. Mitchell (Eds.), Theoretical Aspects of Computer Software. Proceedings, 1994. XI, 887 pages. 1994.

Vol. 790: J. van Leeuwen (Ed.), Graph-Theoretic Concepts in Computer Science. Proceedings, 1993. IX, 431 pages. 1994.

Vol. 791: R. Guerraoui, O. Nierstrasz, M. Riveill (Eds.), Object-Based Distributed Programming. Proceedings, 1993. VII, 262 pages. 1994.

Vol. 792: N. D. Jones, M. Hagiya, M. Sato (Eds.), Logic, Language and Computation. XII, 269 pages. 1994.

Vol. 793: T. A. Gulliver, N. P. Secord (Eds.), Information Theory and Applications. Proceedings, 1993. XI, 394 pages. 1994.

Vol. 794: G. Haring, G. Kotsis (Eds.), Computer Performance Evaluation. Proceedings, 1994. X, 464 pages. 1994.

Vol. 795: W. A. Hunt, Jr., FM8501: A Verified Microprocessor. XIII, 333 pages. 1994.

Vol. 796: W. Gentzsch, U. Harms (Eds.), High-Performance Computing and Networking. Proceedings, 1994, Vol. I. XXI, 453 pages. 1994.

Vol. 797: W. Gentzsch, U. Harms (Eds.), High-Performance Computing and Networking. Proceedings, 1994, Vol. II. XXII, 519 pages. 1994.

Vol. 798: R. Dyckhoff (Ed.), Extensions of Logic Programming. Proceedings, 1993. VIII, 362 pages. 1994.

Vol. 799: M. P. Singh, Multiagent Systems. XXIII, 168 pages. 1994. (Subseries LNAI).

Vol. 800: J.-O. Eklundh (Ed.), Computer Vision – ECCV '94. Proceedings 1994, Vol. I. XVIII, 603 pages. 1994.

Vol. 801: J.-O. Eklundh (Ed.), Computer Vision – ECCV '94. Proceedings 1994, Vol. II. XV, 485 pages. 1994.

Vol. 802: S. Brookes, M. Main, A. Melton, M. Mislove, D. Schmidt (Eds.), Mathematical Foundations of Programming Semantics. Proceedings, 1993. IX, 647 pages. 1994.

Vol. 803: J. W. de Bakker, W.-P. de Roever, G. Rozenberg (Eds.), A Decade of Concurrency. Proceedings, 1993. VII, 683 pages. 1994.

Vol. 804: D. Hernández, Qualitative Representation of Spatial Knowledge. IX, 202 pages. 1994. (Subseries LNAI).

Vol. 805: M. Cosnard, A. Ferreira, J. Peters (Eds.), Parallel and Distributed Computing. Proceedings, 1994. X, 280 pages. 1994.

Vol. 806: H. Barendregt, T. Nipkow (Eds.), Types for Proofs and Programs. VIII, 383 pages. 1994.

Vol. 807: M. Crochemore, D. Gusfield (Eds.), Combinatorial Pattern Matching. Proceedings, 1994. VIII, 326 pages. 1994.

Vol. 808: M. Masuch, L. Pólos (Eds.), Knowledge Representation and Reasoning Under Uncertainty. VII, 237 pages. 1994. (Subseries LNAI).

Vol. 809: R. Anderson (Ed.), Fast Software Encryption. Proceedings, 1993. IX, 223 pages. 1994.

Vol. 810: G. Lakemeyer, B. Nebel (Eds.), Foundations of Knowledge Representation and Reasoning. VIII, 355 pages. 1994. (Subseries LNAI).

Vol. 811: G. Wijers, S. Brinkkemper, T. Wasserman (Eds.), Advanced Information Systems Engineering. Proceedings, 1994. XI, 420 pages. 1994.

Vol. 812: J. Karhumäki, H. Maurer, G. Rozenberg (Eds.), Results and Trends in Theoretical Computer Science. Proceedings, 1994. X, 445 pages. 1994.

Vol. 813: A. Nerode, Yu. N. Matiyasevich (Eds.), Logical Foundations of Computer Science. Proceedings, 1994. IX, 392 pages. 1994.

Vol. 814: A. Bundy (Ed.), Automated Deduction—CADE-12. Proceedings, 1994. XVI, 848 pages. 1994. (Subseries LNAI).

Vol. 815: R. Valette (Ed.), Application and Theory of Petri Nets 1994. Proceedings. IX, 587 pages. 1994.

Vol. 816: J. Heering, K. Meinke, B. Möller, T. Nipkow (Eds.), Higher-Order Algebra, Logic, and Term Rewriting. Proceedings, 1993. VII, 344 pages. 1994.

Vol. 817: C. Halatsis, D. Maritsas, G. Philokyprou, S. Theodoridis (Eds.), PARLE '94. Parallel Architectures and Languages Europe. Proceedings, 1994. XV, 837 pages. 1994.

Vol. 818: D. L. Dill (Ed.), Computer Aided Verification. Proceedings, 1994. IX, 480 pages. 1994.

Vol. 819: W. Litwin, T. Risch (Eds.), Applications of Databases. Proceedings, 1994. XII, 471 pages. 1994.

Vol. 820: S. Abiteboul, E. Shamir (Eds.), Automata, Languages and Programming. Proceedings, 1994. XIII, 644 pages. 1994.

Vol. 821: M. Tokoro, R. Pareschi (Eds.), Object-Oriented Programming. Proceedings, 1994. XI, 535 pages. 1994.

Vol. 822: F. Pfenning (Ed.), Logic Programming and Automated Reasoning. Proceedings, 1994. X, 345 pages. 1994. (Subseries LNAI).

Vol. 823: R. A. Elmasri, V. Kouramajian, B. Thalheim (Eds.), Entity-Relationship Approach — ER '93. Proceedings, 1993. X, 531 pages. 1994.

Vol. 824: E. M. Schmidt, S. Skyum (Eds.), Algorithm Theory – SWAT '94. Proceedings. IX, 383 pages. 1994.

Vol. 825: J. L. Mundy, A. Zisserman, D. Forsyth (Eds.), Applications of Invariance in Computer Vision. Proceedings, 1993. IX, 510 pages. 1994.

Vol. 826: D. S. Bowers (Ed.), Directions in Databases. Proceedings, 1994. X, 234 pages. 1994.

Vol. 827: D. M. Gabbay, H. J. Ohlbach (Eds.), Temporal Logic. Proceedings, 1994. XI, 546 pages. 1994. (Subseries LNAI).

Vol. 828: L. C. Paulson, Isabelle. XVII, 321 pages. 1994.

Vol. 829: A. Chmora, S. B. Wicker (Eds.), Error Control, Cryptology, and Speech Compression. Proceedings, 1993. VIII, 121 pages. 1994.

Vol. 831: V. Bouchitté, M. Morvan (Eds.), Orders, Algorithms, and Applications. Proceedings, 1994. IX, 204 pages. 1994.